IFCoLog Journal of Logic and its Applications

Volume 1, Number 1

June 2014

Disclaimer

Statements of fact and opinion in the articles in IfCoLog Journal of Logics and their Applications are those of the respective authors and contributors and not of the IfCoLog Journal of Logics and their Applications or of College Publications. Neither College Publications nor the IfCoLog Journal of Logics and their Applications make any representation, express or implied, in respect of the accuracy of the material in this journal and cannot accept any legal responsibility or liability for any errors or omissions that may be made. The reader should make his/her own evaluation as to the appropriateness or otherwise of any experimental technique described.

ISBN 978-1-84890-149-0
ISSN (E) 2055-3714
ISSN (P) 2055 3706

College Publications
Scientific Director: Dov Gabbay
Managing Director: Jane Spurr

http://www.collegepublications.co.uk

Printed by Lightning Source, Milton Keynes, UK

SCOPE AND SUBMISSIONS

This journal considers submission in all areas of pure and applied logic, including:

pure logical systems
proof theory
constructive logic
categorical logic
modal and temporal logic
model theory
recursion theory
type theory
nominal theory
nonclassical logics
nonmonotonic logic
numerical and uncertainty reasoning
logic and AI
foundations of logic programming
belief revision
systems of knowledge and belief
logics and semantics of programming
specification and verification
agent theory
databases

dynamic logic
quantum logic
algebraic logic
logic and cognition
probabilistic logic
logic and networks
neuro-logical systems
complexity
argumentation theory
logic and computation
logic and language
logic engineering
knowledge-based systems
automated reasoning
knowledge representation
logic in hardware and VLSI
natural language
concurrent computation
planning

This journal will also consider papers on the application of logic in other subject areas: philosophy, cognitive science, physics etc. provided they have some formal content.

Submissions should be sent to Jane Spurr (jane.spurr@kcl.ac.uk) as a pdf file, preferably compiled in LaTeX using the IFCoLog class file.

CONTENTS

ARTICLES

CONTENTS

ARTICLES

Homogenous and Heterogeneous Logical Proportions

Henri Prade and Gilles Richard
Université Paul Sabatier IRIT

1 Introduction

Commonsense reasoning often relies on the perception of similarity as well as dissimilarity between objects or situations. Such a perception may be expressed and summarized by means of analogical proportions, i.e., statements of the form "A is to B as C is to D". Analogy is not a mere question of similarity between two objects (or situations), but rather a matter of proportion or relation between objects. This view dates back to Aristotle and was enforced by Scholastic philosophy. Indeed, an analogical proportion equates a relation between two objects with the relation between two other objects. As such, the analogical proportion "A is to B as C is to D" poses an analogy of proportionality by (implicitly) stating that the way the two objects A and B, otherwise similar, differ is the same way as the two objects C and D, which are similar in some respects, differ.

A propositional logic modeling of analogical proportions, viewed as a quaternary connective between the Boolean values of some feature pertaining to A, B, C, and D, has been recently proposed in [14]. This logical modeling amounts to precisely state that the difference between A and B is the same as the one between C and D, and that the difference between B and A is the same as the one between D and C. This view can then be proved to be equivalent to state that each time a Boolean feature is true for A and D (resp. A or D) it is also true for B and C (resp. B or C), and conversely. This latter point shows that a counterpart of a characteristic behavior of numerical geometrical proportions ($\frac{a}{b} = \frac{c}{d}$), or of numerical arithmetic proportions ($a - b = c - d$), namely that the product (resp. sum, in the second case) of the extremes is equal to the product (resp. the sum) of the means, is still observed in the logical setting.

However, analogical proportions are not the only type of quaternary statements relying on the ideas of similarity and dissimilarity that can be imagined. They turn out to be a special case of so-called *logical proportions* [17]. Roughly speaking, a logical proportion between four terms A, B, C, D equates similarity or dissimilarity evaluations about the pair (A, B) with similarity or dissimilarity evaluations about the pair (C, D). A set of 120 distinct logical proportions, whose formal expressions share the same structure as well as some remarkable properties, has been identified. Among them, 8 logical proportions stand out as being the only ones that enjoy a code independency property. Namely, their truth status remains unchanged when the truth values 0 and 1 are exchanged. These 8 proportions split into two groups, namely, 4 *homogeneous* ones (which include the analogical proportion) [22], and 4 *heterogeneous* logical proportions, which are dual in some sense of the former ones. The pairs (A, B) and (C, D) play symmetrical roles for homogeneous proportions, while it is not the case for the heterogeneous ones. However, both enjoy noticeable permutation properties.

Similarity and dissimilarity are naturally a matter of degrees. Thus, the extension of homogeneous and heterogeneous logical proportions when features are graded make sense in a multiple-valued logic setting. This makes these logical proportions closer to a symbolic counterpart of numerical proportions where the equality between ratios or differences of quantities may be approximate.

Besides, knowing three values, the statement of the equality of numerical ratios, or of numerical differences, involving a fourth unknown value, and expressing a proportionality relation, is useful for extrapolating this latter value. Similarly, the solving of logical proportion equations may be the basis of reasoning procedures. In particular, when an analogical proportion holds for a large number of features between four situations described by means of n binary features, one may make the plausible inference that the same type of proportion should also hold for a $(n + 1)$th feature. If the truth value of this latter feature is known for three of the situations, and unknown for the fourth one, this value can thus be obtained as the solution of an analogical proportion equation.

The paper is organized as follows. In Section 2, the notion of logical proportions is introduced and formally defined. Then, a structural typology of the different families of logical proportions, as well as some noticeable properties, are presented. Section 3 is devoted to a more detailed study of homogeneous proportions. Section 4 deals with extensions of homogeneous proportions for handling non Boolean or unknown features. This is the case if the features are gradual, or if they are binary but may not apply. It may also happen that for some situations it is not known if a feature holds or not. The section investigates these three types of cases (gradual features, features non applicable, and missing information about a feature), where different multiple-valued logical calculi are involved. Section 5 focuses on heterogeneous proportions, studies their properties, and their extension to gradual properties. Section 6 discusses applications of homogeneous and heterogeneous proportions. Ho-

2

mogeneous logical proportions, especially analogical proportions, seem of interest for completing missing values in tables, a problem sometimes termed "matrix abduction" [1]. It amounts in the logical proportion setting to completing a series A, B, C with X such as (A, B, C, X) makes a proportion of a given type. Heterogeneous logical proportions are shown to be instrumental for picking out the item that does not fit in a list. Thus, the setting of logical proportions appears to be rich enough for coping with two different types of reasoning problems where the ideas of similarity and dissimilarity play a key role in both cases. Psychological quizzes or tests are used for illustrating this ability to exploit comparisons in reasoning.

This paper provides a synthesis of results that have appeared mostly in a series of papers by the authors [19, 18, 22].

2 Logical proportions

Before introducing the formal definitions, let us briefly clarify the notations used.

- When dealing with Boolean logic, $a, b, \ldots$ denote propositional variables (having 0 or 1 as truth value), and we use the standard symbols $\wedge, \vee$ to build up formulas (with parentheses when needed). For the negation operator, instead of using the standard $\neg$ symbol, we will use $\overline{a}$ to denote $\neg a$. This is done for saving space when writing long formulas. As usual $\top$ (resp. $\bot$) denotes the always true (resp. false) proposition.

- 0 and 1 denote the Boolean truth values, and a valuation v is just a function from the set of propositional variables to the set of truth values, i.e., $\{0, 1\}$ in the Boolean case, or $[0, 1]$ in the graded case.

- When we propose a new definition, we will use the symbol $\triangleq$ meaning definitional equality. The right hand side of the equation is the definition of the left hand side.

- When we consider syntactic identity, we use $=_{Id}$: for instance $a \wedge b =_{Id} a \wedge b$ but we do not have $a \wedge b =_{Id} b \wedge a$.

- Finally, the symbol $\equiv$ is reserved for the equivalence, i.e.,

$$a \rightarrow b \triangleq \overline{a} \vee b \qquad a \equiv b \triangleq (a \rightarrow b) \wedge (b \rightarrow a)$$

Logical proportions are Boolean formulas built upon what we called indicators. We introduce this concept in the next subsection and we investigate some fundamental properties.

3

2.1 Similarity and dissimilarity indicators

Generally speaking, the comparison of two items A and B relies on the representation of these items. For instance, the items may be represented as a set of features $\mathcal{A}$ and $\mathcal{B}$. Then, one may define a *similarity measure*. This is the aim of the well-known work of Amos Tversky [26], taking into account the common features, the specificities of A w.r.t. B, and the specificities of B w.r.t. A, respectively modeled by $\mathcal{A} \cap \mathcal{B}$, $\mathcal{A} \setminus \mathcal{B}$, and $\mathcal{B} \setminus \mathcal{A}$. Here, we are not looking for any global measure of similarity, we are rather interested in keeping track in what respect items are similar and in what respect they are dissimilar using Boolean indicators. This is why we adopt a logical setting: features are viewed as Boolean properties. Let P be such a property, which can be seen as a predicate: $P(A)$ may be true (in that case $\neg P(A)$ is false), or false.

When comparing two items A and B w.r.t. such a property P, it makes sense to consider A and B similar (w.r.t. property P):
   - when $P(A) \wedge P(B)$ is true or
   - when $\neg P(A) \wedge \neg P(B)$ is true.
In the remaining cases:
   - when $\neg P(A) \wedge P(B)$ is true or
   - when $P(A) \wedge \neg P(B)$ is true,
we can consider A and B as dissimilar w.r.t. property P.

Since $P(A)$ and $P(B)$ are ground formulas, they can simply be considered as Boolean variables, and denoted a and b by abstracting w.r.t. P. If the conjunction $a \wedge b$ is true, the property is satisfied by both items A and B, while the property is satisfied by neither A nor B if $\overline{a} \wedge \overline{b}$ is true. The property is true for A only (resp. B only) if $a \wedge \overline{b}$ (resp. $\overline{a} \wedge b$) is true. This is why we call such a conjunction of Boolean literals an *indicator*, and for a given pair of Boolean variables (a, b), we have exactly 4 distinct indicators:

   • $a \wedge b$ and $\overline{a} \wedge \overline{b}$ that we call *similarity indicators*,

   • $a \wedge \overline{b}$ and $\overline{a} \wedge b$ that we call *dissimilarity indicators*.

Let us observe that negating anyone of the two terms of a dissimilarity indicator turns it into a similarity indicator, and conversely. Hence, negating the two terms of an indicator yields an indicator of the same type.

2.2 Building logical proportions with indicators

When describing two elementary situations encoded by two Boolean variables a and b, one may use one of the four above indicators. Putting such a description in relation with what takes place with two other Boolean variables c and d in terms of some indicator, leads to state an equivalence between one indicator pertaining to the pair (a, b) and one indicator pertaining to the pair (c, d). However, one may consider that using *two* indicators to describe the status of 2 variables a and b may be more satisfactory from some symmetrization point of view than using only one indicator. For instance, using $\overline{a} \wedge b$ together with $a \wedge \overline{b}$ establishes the symmetry between a and b, or using $a \wedge \overline{b}$ together with $a \wedge b$ considers counter-examples as well as examples in context a, or using $\overline{a} \wedge \overline{b}$ together with $a \wedge b$ provides the same role to negative or positive features. Note that such symmetrizations occur for free with numerical proportions where for instance one can exchange a and b on the one hand, c and d on the other hand, still writing a unique equality. It is why we more particularly focus on proportions defined as the conjunction of *two distinct* equivalences between an indicator for the pair (a, b) and an indicator for the pair (c, d).

One may wonder about the simultaneous use of three indicators for comparing two Boolean variables. This would lead to three equivalences instead of two, which appears conceptually more complicated, and maybe farther from the idea of proportion inherited from the numerical setting. Then, for the sake of simplicity, we stick to the conjunctions of two equivalences between indicators in the following. This defines a so-called *logical proportion* [17, 19]. More formally, let us denote $I_{(a,b)}$ and $I'_{(a,b)}$[1] (resp. $I_{(c,d)}$ and $I'_{(c,d)}$) 2 indicators for (a, b) (resp. (c, d)). Then

Definition 1. *A logical proportion $T(a, b, c, d)$ is the conjunction of 2 distinct equivalences between indicators of the form*

$$I_{(a,b)} \equiv I_{(c,d)} \wedge I'_{(a,b)} \equiv I'_{(c,d)}$$

An example of such proportion is $((\overline{a} \wedge \overline{b}) \equiv (c \wedge \overline{d})) \wedge ((\overline{a} \wedge b) \equiv (\overline{c} \wedge d))$ where

- $I_{(a,b)} \triangleq \overline{a} \wedge \overline{b}, \quad I_{(c,d)} \triangleq c \wedge \overline{d},$
- $I'_{(a,b)} \triangleq \overline{a} \wedge b, \quad I'_{(c,d)} \triangleq \overline{c} \wedge d.$

Obviously, this formal definition goes beyond what may be expected from the informal idea of "logical proportion", since equivalences may be put between things that are not homogeneous (i.e., mixing similarity and dissimilarity indicators in various ways).

Let us first determine the number of logical proportions. To build an equivalence between indicators, we have to choose one indicator among four for the pair (a, b)

---

[1]Note that $I_{(a,b)}$ (or $I'_{(a,b)}$) refers to one element in the set $\{a \wedge b, \overline{a} \wedge b, a \wedge \overline{b}, \overline{a} \wedge \overline{b}\}$, and should not be considered as a functional symbol. Still, we use this notation for the sake of readability.

and similarly for the pair (c, d), we get $4 \times 4 = 16$ distinct equivalences. To build up a logical proportion, we first choose one equivalence among 16, and then the second equivalence has to be chosen among the 15 remaining ones, leading to $16 \times 15 = 240$ pairs of equivalences. Taking into account the commutativity of the Boolean conjunction, we finally get $240/2 = 120$ potentially distinct logical proportions. We shall see in subsection 2.4 that they are indeed distinct. We first provide a syntactic typology of the logical proportions.

2.3 Typology of logical proportions

Logical proportions can be classified according to the ways they are built up. At this stage, it makes sense to distinguish between two types of indicators: similarity indicators that are denoted by S, and dissimilarity indicators that are denoted by D: e.g., $D_{(a,b)} \in \{a \wedge \overline{b}, \overline{a} \wedge b\}$.

Depending on the way the indicators are chosen, one may mix the similarity and the dissimilarity indicators differently in the definition of a proportion.

This leads us to distinguish a specific subfamily of proportions, the so-called *degenerated proportions*: those ones involving only 3 distinct indicators in their definition. For instance

$$(a \wedge b \equiv \overline{c} \wedge d) \wedge (\overline{a} \wedge b \equiv \overline{c} \wedge d)$$

is such a proportion where $I_{(c,d)} =_{Id} I'_{(c,d)}$.

For the remaining proportions, it is required that all the indicators appearing in the definition of the proportion are distinct. At this stage, among the non-degenerated proportions, we can identify 4 subfamilies that we describe below:

- **The 4 homogeneous proportions**

  For these proportions, we do not mix different types of indicators in the 2 equivalences. The homogeneous proportions are of the form

  $$S_{(a,b)} \equiv S_{(c,d)} \wedge S'_{(a,b)} \equiv S'_{(c,d)}$$

  or

  $$D_{(a,b)} \equiv D_{(c,d)} \wedge D'_{(a,b)} \equiv D'_{(c,d)}$$

  Thus, it appears that only 4 proportions among 120 are homogeneous. They are (with their name):

  – *analogy* : $A(a, b, c, d)$, defined by

  $$((a \wedge \overline{b}) \equiv (c \wedge \overline{d})) \wedge ((\overline{a} \wedge b) \equiv (\overline{c} \wedge d))$$

6

- *reverse analogy*: $R(a, b, c, d)$, defined by

$$((a \wedge \overline{b}) \equiv (\overline{c} \wedge d)) \wedge ((\overline{a} \wedge b) \equiv (c \wedge \overline{d}))$$

- *paralogy* : $P(a, b, c, d)$, defined by

$$((a \wedge b) \equiv (c \wedge d)) \wedge ((\overline{a} \wedge \overline{b}) \equiv (\overline{c} \wedge \overline{d}))$$

- *inverse paralogy*: $I(a, b, c, d)$, defined by

$$((a \wedge b) \equiv (\overline{c} \wedge \overline{d})) \wedge ((\overline{a} \wedge \overline{b}) \equiv (c \wedge d))$$

Analogy already appeared under this form in [14]; paralogy and reverse analogy were first introduced in [16], and inverse paralogy in [19]. While the analogical proportion (analogy, for short) reads "*a* is to *b* as *c* is to *d*" and expresses that "*a* differs from *b* as *c* differs from *d*, and conversely *b* differs from *a* as *d* differs from *c*", reverse analogy expresses that "*a* differs from *b* as *d* differs from *c*, and conversely *b* differs from *a* as *c* differs from *d*", paralogy expresses that "what *a* and *b* have in common, *c* and *d* have it also" (positively and negatively). Paralogy is a given name. Finally, *inverse paralogy* expresses that "what *a* and *b* have in common, *c* and *d* miss it, and conversely". As can be seen, inverse paralogy expresses a form of antinomy between pairs (a, b) and (c, d). Note that we use two different words, "inverse" and "reverse", since the changes between analogy and reverse analogy on the one hand, and paralogy and inverse paralogy on the other hand, are not of the same nature. From now on, we denote analogy with A, reverse analogy with R, paralogy with P, inverse analogy with I. When we need to denote any unspecified proportion, we will use the letter T.

- **The 16 conditional proportions**

  Their expression is made of the conjunction of an equivalence between similarity indicators and of an equivalence between dissimilarity indicators. Thus, they are of the form

  $$S_{(a,b)} \equiv S_{(c,d)} \wedge D_{(a,b)} \equiv D_{(c,d)}$$

  There are 16 conditional proportions (2×2 choices *per* equivalence). An example is

  $$((a \wedge b) \equiv (c \wedge d)) \wedge ((a \wedge \overline{b}) \equiv (c \wedge \overline{d}))$$

Let us explain the term "conditional". It comes from the fact that these proportions express "equivalences" between conditional statements. Indeed, it has been advocated in [5] that a rule "if a then b" can be seen as a three valued entity that is called 'conditional object' and denoted $b|a$ [4]. This entity is:

- true if $a \wedge b$ is true. The elements making it true are the examples of the rule "if a then b",
- false if $a \wedge \overline{b}$ is true. The elements making it true are the counter-examples of the rule "if a then b",
- undefined if $\overline{a}$ is true. The rule "if a then b" is then not applicable.

Thus, the above proportion $((a \wedge b) \equiv (c \wedge d)) \wedge ((a \wedge \overline{b}) \equiv (c \wedge \overline{d}))$ may be denoted $b|a :: d|c$ combining the two conditional objects in the spirit of the usual notation for analogical proportion. Indeed, it expresses a semantical equivalence between the 2 rules "if a then b" and "if c then d" by stating that they have the same examples, i.e. $(a \wedge b) \equiv (c \wedge d)$) and the same counter-examples $(a \wedge \overline{b}) \equiv (c \wedge \overline{d})$.

It is worth noticing that such proportions have equivalent forms, e.g.:

$$(b|a :: d|c) \equiv (\overline{b}|a :: \overline{d}|c)$$

which agrees with the above semantics and more generally with the idea of conditioning. Indeed the examples "if a then b" are the counter-examples of "if a then $\overline{b}$", and vice-versa. Due to this remark, it is enough to consider the equivalences between one of the 4 conditional objects $a|b$, $b|a$, $a|\overline{b}$, $b|\overline{a}$, and the 4 other conditional objects built with (c, d), yielding 4×4 proportions as expected. Besides, 8 conditional proportions have been first considered in [19], but not the 8 remaining ones, since they do not satisfy the "full identity" property, discussed in the next section.

- **The 20 hybrid proportions**

  They are characterized by equivalences between similarity and dissimilarity indicators in their definitions. They are of the form.

$$S_{(a,b)} \equiv D_{(c,d)} \wedge S'_{(a,b)} \equiv D'_{(c,d)}$$

or

$$D_{(a,b)} \equiv S_{(c,d)} \wedge D'_{(a,b)} \equiv S'_{(c,d)}$$

8

or

$$S_{(a,b)} \equiv D_{(c,d)} \wedge D_{(a,b)} \equiv S_{(c,d)}.$$

There are 20 hybrid proportions: 2 of the first type, 2 of the second type, 16 of the third type since we have here 4 choices for an equivalence $S_{(a,b)} \equiv D_{(c,d)}$, and 4 choices for $D_{(a,b)} \equiv S_{(c,d)}$.

If we remember that negating anyone of the two terms of a dissimilarity indicator turns it into a similarity indicator, and conversely, we understand that changing a into $\overline{a}$ (and $\overline{a}$ into a), or applying a similar transformation with respect to b, c, or d, turns

- an hybrid proportion into an homogeneous or a conditional proportion;

- an homogeneous or a conditional proportion into an hybrid proportion.

This indicates the close relationship of hybrid proportions with homogeneous and conditional proportions. More precisely,

- on the one hand there are 4 hybrid proportions such that replacing a with $\overline{a}$ leads to the 4 homogeneous proportions A, R, P, I. They are obtained by the two first kinds of patterns for building hybrid proportions. Moreover, we shall see in the next section that they constitute with the 4 homogeneous proportions the 8 proportions that are the only ones satisfying "code independency" property.

- on the other hand, there are 16 remaining hybrid proportions, obtained by the third kind of pattern for building them. They can be written as the equivalence of 2 conditional objects, although they do not obey the conditional proportion pattern. For instance, $((\overline{a} \wedge b) \equiv (c \wedge d)) \wedge ((a \wedge b) \equiv (\overline{c} \wedge d))$ can be written as $\overline{a}|b :: c|d$. This proportion is indeed obtained from the conditional proportion $a|b :: c|d$ by changing a into $\overline{a}$. Thus, these 16 new equivalences between conditional objects are not of the form $a|b :: c|d$ (or equivalently $\overline{a}|b :: \overline{c}|d$) produced by the pattern of conditional proportions, but of a "mixed" form having an odd number of negated terms.

- **The 32 semi-hybrid proportions**

One half of their expressions involve indicators of the same type, while the other half requires equivalence between indicators of opposite types. They are of the form

9

$$S_{(a,b)} \equiv S_{(c,d)} \wedge S'_{(a,b)} \equiv D_{(c,d)}$$

or

$$S_{(a,b)} \equiv S_{(c,d)} \wedge D_{(a,b)} \equiv S'_{(c,d)}$$

or

$$D_{(a,b)} \equiv D_{(c,d)} \wedge S_{(a,b)} \equiv D'_{(c,d)}$$

or

$$D_{(a,b)} \equiv D_{(c,d)} \wedge D'_{(a,b)} \equiv S_{(c,d)}$$

There are 32 semi-hybrid proportions (8 of each kind: 4 choices for the first equivalence, times 2 choices for the element that is not of the same type as the three others (D or S) in the second equivalence). An example of semi-hybrid proportion is $((a \wedge b) \equiv (c \wedge d)) \wedge ((\bar{a} \wedge \bar{b} \equiv (\bar{c} \wedge d))$.

Applying a change from a to $\bar{a}$ (and $\bar{a}$ to a), or applying a similar transformation with respect to b, c, or d, turns a semi-hybrid proportion into a semi-hybrid proportion (since as already said, negating anyone of the two terms of a dissimilarity indicator turns it into a similarity indicator, and conversely). This contrasts with the hybrid proportion class which is not closed under such a transformation.

- **The 48 degenerated proportions**

  In all the above categories, the 4 indicators related by equivalence symbols should be all distinct. In degenerated proportions, there are only 3 different indicators and it is simpler to come back to our initial notation. With this notation, these proportions are of the form

  $$I_{(a,b)} \equiv I_{(c,d)} \wedge I_{(a,b)} \equiv I'_{(c,d)}$$

  or

  $$I_{(a,b)} \equiv I_{(c,d)} \wedge I'_{(a,b)} \equiv I_{(c,d)}$$

  Their number is easy to compute: we have to choose $I_{(a,b)}$ among 4 indicators and then to choose 2 distinct indicators among 4 pertaining to (c,d): we then get 4 * 6 = 24 proportions of the first form. The same reasoning with the second kind of expression leads to a total of 48 degenerated proportions. Note that the change from a to $\bar{a}$ (and $\bar{a}$ to a), or a similar transformation with respect to b, c, or d, turns a degenerated proportion into a degenerated proportion.

It can be seen that degenerated proportions always involve a mutual exclusiveness condition between 2 positive or negative literals pertaining to either the pair (a, b) or the pair (c, d). Indeed, if we consider the first form, we get $I_{(a,b)} \equiv I_{(c,d)}$ on the one hand, and $I_{(c,d)} \equiv I'_{(c,d)}$ on the other hand, i.e. an equivalence between two syntactically distinct indicators pertaining to the same pair (c, d). There are 6 cases only:

- $(\overline{c} \wedge d) \equiv (c \wedge \overline{d})$ iff $c \equiv d$
- $(c \wedge d) \equiv (\overline{c} \wedge \overline{d})$ iff $c \equiv \overline{d}$
- $(c \wedge d) \equiv (c \wedge \overline{d})$ iff $c \equiv \bot$
- $(c \wedge d) \equiv (\overline{c} \wedge d)$ iff $d \equiv \bot$
- $(\overline{c} \wedge d) \equiv (\overline{c} \wedge \overline{d})$ iff $\overline{c} \equiv \bot$
- $(c \wedge \overline{d}) \equiv (\overline{c} \wedge \overline{d})$ iff $\overline{d} \equiv \bot$

Thus, we also have $I_{(a,b)} \equiv \bot$ (since we have $I_{(c,d)} \equiv \bot$ and $I'_{(c,d)} \equiv \bot$), which expresses a mutual exclusiveness condition. Since we have 4 possible choices for $I_{(a,b)}$, it yields $4 \times 6 = 24$ distinct proportions, and exchanging (a, b) with (c, d) gives the 24 other degenerated proportions. Generally speaking, degenerated proportions correspond to a mutual exclusiveness condition between component(s) or negation of component(s) of one of the pairs (a, b) or (c, d), together with

- either an identity condition pertaining to the other pair,

- or a tautology condition on one of the literals of the other pair without any constraint on the other literal.

2.4   Basic properties of logical proportions

In this subsection, we first establish a remarkable property that single out the logical proportion s among the whole set of quaternary Boolean formulas. In order to do that we need a lemma.

Lemma 1. *An equivalence between indicators has exactly 10 valid valuations.*

Proof: Such an equivalence $eq \triangleq I_{a,b} \equiv I_{c,d}$ is satisfied only when it matches one of the 2 patterns $1 = 1$ or $0 = 0$: due to the fact that 0 is an absorbing value for $\wedge$, these patterns correspond to the 10 valuations shown in Table 1 for the literals

11

involved in the indicators (with obvious notation). Any other valuation[2] does not match anyone of the 2 previous patterns and will lead to the truth value 0 for the equivalence eq. □

Table 1: 10 valid valuations for an equivalence between indicators

literal 1	literal 2	literal 3	literal 4	pattern
1	1	1	1	$1 = 1$
0	1	0	1	$0 = 0$
0	1	1	0	$0 = 0$
0	1	0	0	$0 = 0$
1	0	0	1	$0 = 0$
1	0	1	0	$0 = 0$
1	0	0	0	$0 = 0$
0	0	0	1	$0 = 0$
0	0	1	0	$0 = 0$
0	0	0	0	$0 = 0$

Proposition 1. *The truth table of a logical proportion has 6 and only 6 valuations with truth value 1.*

Proof: Since a logical proportion T is the conjunction $eq_1 \wedge eq_2$ of 2 equalities between indicators, with $eq_1 \neq eq_2$, it appears from Lemma 1 that T has a maximum of 10 valid valuations and a minimum of 4 valid valuations. Let us start from $eq1$, having 10 valid valuations which are candidate to validate T. Obviously, adding eq_2 to eq_1 will reduce the number of valid valuations for T. Let us assume eq_2 differs from eq_1 with only one literal (or negation operator). This is then a degenerated proportion. Without loss of generality, we can consider that the difference between eq_1 and eq_2 occurs on the first literal meaning eq_1 is $a \wedge l_2 \equiv l_3 \wedge l_4$ and eq_2 is $\overline{a} \wedge l_2 \equiv l_3 \wedge l_4$ or vice versa. It is then quite clear that the first valuation 1111 valid for eq_1 is not valid any more for T. It remains 9 candidates valuations. Finally any valuation starting with 01 is not valid any more and we have 3 such valuations. All the 6 remaining valuations are still valid for T. Which ends the proof when the 2 equalities differ from one negation (i.e. one literal). Now when they differ from 2 literals, two cases have to be considered:

- either the 2 literals where eq_1 differs from eq_2 are on the same side of an equivalence i.e. eq_2 is $l'_1 \wedge l'_2 \equiv l_3 \wedge l_4$ (degenerated proportion)

---

[2]The only valuations considered in this paper pertain to 4-tuples of variables. In practice, a Boolean valuation v will be denoted by the values $v(a)v(b)v(c)v(d)$ without any blank space, e.g., 0100 is short for $v(a) = 0, v(b) = 1, v(c) = 0, v(d) = 0$.

- or they are on different side i.e. eq_2 is $l'_1 \wedge l_2 \equiv l'_3 \wedge l_4$.

In the first case, the valuations $1111, 0010, 0001$ and 0000 are not valid any more, but all other ones remain valid. In the second case, the valuations $0100, 0110, 1001$ and 0001 are not valid anymore, but all the other ones remain valid. We are done for the case of 2 differences. When they differ from 3 literals, let us suppose l_4 appears in both equivalence, the valuations $1001, 0101, 0010$ and 0000 are not valid anymore and we stick with the 6 remaining ones. In the case where all the literals are different, obviously the 4 valuations containing only one occurrence of 1 are not valid anymore because they lead to an invalid pattern 0=1 or 1=0 for eq_2. And we have exactly 4 such valuations. It remains 6 valid valuations. □

Note that the negation of a logical proportion is not a logical proportion since such a negation has 10 valuations leading to true in its table. Besides, the 120 logical proportions are all distinct as shown below with the help of the following lemma.

Lemma 2. *Two equivalences between indicators have the same truth table iff they are identical.*

Proof. It is sufficient to show that if 2 equalities eq_1 and eq_2 have the same truth table, then they are syntactically identical. In other terms, we have to prove that $eq_1 \equiv eq_2$ implies $eq_1 =_{Id} eq_2$. Without loss of generality, let us assume that eq_1 contains a but eq_2 contains $\overline{a}$. Considering the unique valuation v such that $v(eq_1) = 1$ with the pattern $1 = 1$, v is such that $v(a) = 1$. By hypothesis, $v(eq_2) = 1$ but in that case with the pattern $0 = 0$ since $v(\overline{a}) = 0$. Let us now modify v into v' such that $v'(a) = v(a) = 0, v'(c) = v(c), v'(d) = v(d)$ and $v'(b) = v(b)$. Obviously v' does not validate eq_1 but validates eq_2 which contradicts the hypothesis. □

Proposition 2. *The truth tables of the 120 proportions are all distinct.*

Proof. We are going to show that, when 2 proportions $T \triangleq eq_1 \wedge eq_2$ and $T' \triangleq eq'_1 \wedge eq'_2$ have the same truth table, they are syntactically identical (up to a permutation of the 2 equalities). In other words, $T \equiv T'$ implies $T =_{Id} T'$. Starting from $T \equiv T'$, it amounts to show that if eq_1 is syntactically different from eq'_1, eq_1 is syntactically equal to eq'_2. This will complete the proof as a similar reasoning will show that eq_2 is, in the same context, syntactically equal to eq'_1.

In fact, if eq_1 is syntactically different from eq'_1, we can assume for instance without loss of generality that eq_1 contains a but eq'_1 contains $\overline{a}$. Let us consider the unique valuation σ, validating T and T', such that $\sigma(eq_1) = 1$ with the pattern $1 = 1$. Necessarily, this valuation σ is such that $\sigma(a) = 1$. By hypothesis, $\sigma(eq'_1) = 1$ but in that case with the pattern $0 = 0$ since $\sigma(\overline{a}) = 0$. Let us now modify σ into σ'

such that $\sigma'(a) = \sigma(a) = 0, \sigma'(c) = \sigma(c), \sigma'(d) = \sigma(d)$ and $\sigma'(b) = \sigma(b)$. Obviously $\sigma'(T) = \sigma'(eq_1) = 0$ but $\sigma'(eq_1) = 1$ still following the pattern $0 = 0$. The only option for having $\sigma(T) = \sigma(T') = 0$ is thus to have $\sigma'(eq_2') = 0$ which means a belongs to eq_2'. Continuing the same reasoning, we show that $eq_1 =_{Id} eq_2'$ and we infer that if $eq_1 \neq eq_1'$, necessarily $eq_1 =_{Id} eq_2$. $\qquad\square$

Combined with the fact that there are $C_{16}^6 = 8008$ truth tables with 16 lines, this result makes logical proportions quite rare in the world of quaternary Boolean formulas.

An exhaustive investigation of the whole set of logical proportions with respect to various other properties has been done in [19, 22, 21]. In the next subsection, we focus on one of these properties which allows us to characterize a small subset of remarkable proportions.

2.5 Code independency

Just as a numerical proportion holds independently of the base used for encoding numbers, or of the system of units representing the quantities at hand, it seems desirable that a logical proportion should be independent of the way we encode items in terms of the truth or the falsity of features. It means that the formula defining a proportion T should be valid when we switch 0 to 1 and 1 to 0. The formal expression of this property, that we call *code independency*, writes:

$$T(a, b, c, d) \rightarrow T(\overline{a}, \overline{b}, \overline{c}, \overline{d})$$

Surprisingly, this property highlights the fact once more that a single equivalence would not lead to a satisfactory definition for a logical proportion. Indeed, a unique equivalence between indicators, denoted $l_1 \wedge l_2 \equiv l_3 \wedge l_4$, where the l_i's are literals does not satisfy *code independency*, as explained now. If we consider a valuation v such that $v(l_1) = v(l_2) = v(l_3) = 0$ and $v(l_4) = 1$, obviously v makes the equivalence valid since $v(l_1 \wedge l_2) = v(l_3 \wedge l_4) = 0$. But when we switch 0 to 1 and 1 to 0, it appears that the new valuation v' such that $v'(l_1) = v'(l_2) = v'(l_3) = 1$ and $v'(l_4) = 0$ does not validate the equivalence anymore. This shows that one equivalence is not enough if we are interested in "code independency". We have to consider at least 2 equivalences to capture this behavior. For instance, $(a \wedge b \equiv c \wedge d) \wedge (\overline{a} \wedge \overline{b} \equiv \overline{c} \wedge \overline{d})$ clearly satisfies code independency.

Unfortunately, being built as the conjunction of two equivalences is not a sufficient condition for code independency, and many logical proportions do not satisfy it. We have the following result:

Proposition 3. *There are exactly 8 proportions satisfying the code independency property: the 4 homogeneous proportions A, R, P, I, and 4 hybrid proportions (shown in Table 2).*

Proof: In fact, the code independency property implies a complete equivalence:

$$T(a, b, c, d) \leftrightarrow T(\overline{a}, \overline{b}, \overline{c}, \overline{d})$$

Since both $T(a, b, c, d)$ and $T(\overline{a}, \overline{b}, \overline{c}, \overline{d})$ are logical proportions, Proposition 2 tells us that the 2 proportions should be identical up to a permutation of the 2 equalities. This exactly means that the second equivalence is obtained from the first one by negating all the variables. Since we have 4×4 equalities between indicators, we can build exactly $16/2 = 8$ proportions satisfying code independency property: each time we choose an equivalence, we use it and its negated form to build up a suitable proportion. Since A, R, P, I are built this way, they satisfy code independency. □

Table 2: The 4 hybrid proportions satisfying code independency

$\mathbf{H_a}$	$\mathbf{H_b}$
$(\overline{a} \wedge b \equiv \overline{c} \wedge \overline{d}) \wedge (a \wedge \overline{b} \equiv c \wedge d)$	$(\overline{a} \wedge b \equiv c \wedge \overline{d}) \wedge (a \wedge \overline{b} \equiv \overline{c} \wedge d)$
$\mathbf{H_c}$	$\mathbf{H_d}$
$(\overline{a} \wedge \overline{b} \equiv \overline{c} \wedge d) \wedge (a \wedge b \equiv c \wedge \overline{d})$	$(\overline{a} \wedge \overline{b} \equiv c \wedge d) \wedge (a \wedge b \equiv \overline{c} \wedge \overline{d})$

As a consequence of this result, this set of 8 proportions stand out of the whole set of 120 proportions. This set of proportions is clearly divided in 2 subsets: the 4 homogeneous proportions on one hand, and the 4 remaining ones, that we call *heterogeneous* proportions, on the other hand. In the next two sections, we first investigate the 4 homogeneous proportions through the angle of a list of meaningful properties, as well as their interrelationships, and their extensions to multiple-valued settings. After which, we shall move to the study of the 4 heterogeneous proportions in Section 5.

3   The 4 homogeneous proportions

We investigate now the 4 homogeneous proportions A, R, P, I from a semantical point of view. When considered as Boolean formulas, their semantics is given via their truth tables (which have $2^4 = 16$ lines since these proportions involve 4 variables).

3.1 Boolean truth tables

Starting from their syntactic expressions, it is an easy game to build up the truth tables of proportions A, R, P, I: they are exhibited in Table 3, where only the valuations leading to the truth value 1, are shown. This means that all the other ones lead to the truth value 0. As expected, only 6 valuations among 16 in the tables lead to a truth value 1. We also observe that there are only 8 distinct valuations that appear in Table 3. This emphasizes their collective coherence as the whole class of homogeneous proportions. Moreover, they go by pairs where 0 and 1 are exchanged, thus pointing out their "code independency".

Table 3: Analogy, Reverse analogy, Paralogy, Inverse paralogy truth tables

A				R				P				I			
0	0	0	0	0	0	0	0	0	0	0	0	1	1	0	0
1	1	1	1	1	1	1	1	1	1	1	1	0	0	1	1
0	0	1	1	0	0	1	1	1	0	0	1	1	0	0	1
1	1	0	0	1	1	0	0	0	1	1	0	0	1	1	0
0	1	0	1	0	1	1	0	0	1	0	1	0	1	0	1
1	0	1	0	1	0	0	1	1	0	1	0	1	0	1	0

It is interesting to take a closer look at the truth tables of the four homogeneous proportions. First, one can observe in Table 3, that 8 possible valuations for (a, b, c, d) never appear among the patterns that make A, R, P, or I true: these 8 valuations are of the form $xxxy, xxyx, xyxx$, or $yxxx$ with $x \neq y$ and $(x, y) \in \{0, 1\}^2$. As can be seen, it corresponds to situations where $a = b$ and $c \neq d$, or $a \neq b$ and $c = d$, i.e., similarity holds between the components of one of the pairs, and dissimilarity holds in the other pair. Moreover, the truth table of each of the four homogeneous proportions, is built in the same manner:

1. 2 lines of the table correspond to the characteristic pattern of the proportion; namely the two lines where one of the two equivalences in its definition holds true under the form $1 \equiv 1$ (rather than $0 \equiv 0$). Thus,

   - A is characterized by the pattern $xyxy$ (corresponding to valuations 1010 and 0101), i.e. we have the same difference between a and b as between c and d;

   - R is characterized by the pattern $xyyx$ (corresponding to valuations 1001 and 0110), i.e., the differences between a and b and between c and d are in opposite directions;

   - P is characterized by the pattern $xxxx$ (corresponding to valuations 1111 and 0000), i.e., what a and b have in common, c and d have it also;

16

- I is characterized by the pattern $xxyy$ (corresponding to valuations 1100 and 0011), i.e. what a and b have in common, c and d do not have it, and conversely.

2. the 4 other lines of the truth table of an homogeneous proportion T are generated by the characteristic patterns of the two other proportions that are not opposed to T (in the sense that A and R are opposed, as well as P and I). For these four lines, the proportion holds true since its expression reduces to $(0 \equiv 0) \wedge (0 \equiv 0)$.

Thus, the six lines of the truth table of A that makes it true are induced by the characteristic patterns of A, P, and I[3], the six valuations that makes P true are induced by the characteristic patterns of P, A, and R, and so on for R and I.

3.2 Relevant properties

Before going deeper in the investigation, remember that the Boolean analogical proportion is supposed to be, in a Boolean setting, the counterpart of the classical numerical proportions. Then, it is interesting to consider Boolean counterparts of the properties satisfied by the numerical proportions, other than *code independency*. We list these properties below (with T denoting a logical proportion).

- *Full identity*: A numerical proportion holds when all the numbers are equal, i.e., $a = b = c = d$, which logically translates into

$$T(a, a, a, a)$$

- *Reflexivity*: A numerical proportion holds between (a, b) and (a, b) which logically translates into
$$T(a, b, a, b)$$

Obviously, *reflexivity* entails *full identity*.

- *Sameness*: A numerical proportion holds between (a, a) and (b, b), which logically translates into
$$T(a, a, b, b)$$

Still, *sameness* entails *full identity*.

---

[3]The measure of analogical dissimilarity introduced in [13] is 0 for the valuations corresponding to the characteristic patterns of A, P, and I, maximal for the valuations corresponding to the characteristic patterns of R, and takes the same intermediary value for the 8 valuations characterized by one of the patterns $xxxy$, $xxyx$, $xyxx$, or $yxxx$.

- *symmetry* : We can exchange the pair (a, b) with the pair (c, d) in the numerical proportion, which logically translates into

$$T(a, b, c, d) \rightarrow T(c, d, a, b)$$

- *Central (and extreme) permutation* : This is a well known property of numerical proportions, which logically translates into

$$T(a, b, c, d) \rightarrow T(a, c, b, d) \text{ (central permutation)}$$

and

$$T(a, b, c, d) \rightarrow T(d, b, c, a) \text{ (extreme permutation)}$$

- *Transitivity*: This property that holds for numerical proportions is logically stated as follows

$$T(a, b, c, d) \wedge T(c, d, e, f) \rightarrow T(a, b, e, f)$$

- *Exchange-mirroring*: The negation operator can play for Boolean values the role of an inverse operator for numbers. A numerical proportion holds between a pair (a, b) and the pair(b^{-1}, a^{-1}), which logically translates into

$$T(a, b, \overline{b}, \overline{a})$$

- *Semi-mirroring*: Similarly it is worth to consider

$$T(a, b, \overline{a}, \overline{b})$$

This property is not satisfied by numerical proportions.

- *Negation-compatibility*: Similarly it is worth to consider

$$T(a, \overline{a}, b, \overline{b})$$

This property is also not satisfied by numerical proportions.

Investigating the homogeneous proportions with regard to the properties listed above can simply be done with an examination of the truth table of the target proportion. We summarize in Table 4 all the properties satisfied by A, R, P, I: the third column enumerates the homogeneous proportions satisfying the property, respectively named and described in the 1st and 2nd columns.

Table 4: Boolean properties of A, R, P, I

Property name	Formal definition	Proportion
full identity	$T(a,a,a,a)$	A,R,P
reflexivity	$T(a,b,a,b)$	A,P
reverse reflexivity	$T(a,b,b,a)$	R,P
sameness	$T(a,a,b,b)$	A,R
symmetry	$T(a,b,c,d) \to T(c,d,a,b)$	A,R,P,I
permutation of means	$T(a,b,c,d) \to T(a,c,b,d)$	A,I
permutation of extremes	$T(a,b,c,d) \to T(d,b,c,a)$	A,I
all permutations	$\forall i,j, T(a,b,c,d) \to T(p_{i,j}(a,b,c,d))$	I
transitivity	$T(a,b,c,d) \land T(c,d,e,f) \to T(a,b,e,f)$	A,P
semi-mirroring	$T(a,b,\overline{a},\overline{b})$	R,I
exchange mirroring	$T(a,b,\overline{b},\overline{a})$	A,I
negation compatib.	$T(a,\overline{a},b,\overline{b})$	P,I

Note that the 4 homogeneous proportions satisfy symmetry: $T(a,b,c,d) = T(c,d,a,b)$, as well as many other properties. In particular, analogical proportion A enjoys properties that parallel properties of numerical proportions: full identity, reflexivity, symmetry, central and extreme permutations, and transitivity.

One can also establish properties linking the homogeneous proportions, which are easily deducible from their definitions in terms of indicators.

Proposition 4.
$A(a,b,c,d) \equiv R(a,b,d,c);\ A(a,b,c,d) \equiv P(a,d,c,b);\ A(a,b,c,d) \equiv I(\overline{a},d,\overline{c},b)$

As can be seen, homogeneous proportions are strongly linked together. Especially A, R, P are exchanged through simple permutation s; in that respect, I stands apart. Besides, A, R, P, I are mutually exclusive, as a simple examination of their truth tables reveals that their intersection is empty.

Proposition 5. $A(a,b,c,d) \land R(a,b,c,d) \land P(a,b,c,d) \land I(a,b,c,d) = \bot$

Lastly, having a closer look on the homogeneous proportions, we can easily build Table 5 which gives what $T(a,b,c,d) \land T(c,d,e,f)$ entails for the 4 homogeneous proportions.

All these common properties explain why the homogeneous proportions stand out from the whole set of 120 logical proportions. It makes homogeneous proportions a worth considering Boolean counterpart of numerical proportions.

Table 5: Chaining properties for A, R, P, I

chaining	result	transitivity
$A \wedge A$	A	yes
$R \wedge R$	A	no
$P \wedge P$	P	yes
$I \wedge I$	P	no
$A \wedge R$	R	
$P \wedge I$	I	

3.3   Characterization of homogeneous proportions by properties

Some subsets of the properties listed above are sufficient for characterizing one or more homogeneous proportions as unique among the 120 logical proportions. Let us start with the following result:

Proposition 6.   • A, R, P are the unique proportions to satisfy full identity and code independency.

- A is the only proportion to satisfy sameness $(T(a, a, b, b))$ and reflexivity $(T(a, b, a, b))$.

- R is the only proportion to satisfy sameness and reverse reflexivity $T(a, b, b, a)$.

- P is the only proportion to satisfy reflexivity and reverse reflexivity.

- There is no proportion simultaneously satisfying sameness, reflexivity, and reverse reflexivity.

Proof: The first statement comes from Proposition 3 giving the 8 proportions satisfying code independency, along with an immediate checking of the proportions syntactic form. For instance, H_a defined as $(a \wedge b \equiv c \wedge \overline{d}) \wedge (\overline{a} \wedge \overline{b} \equiv \overline{c} \wedge d)$ is definitely not valid for valuation 0000. The same reasoning applies to all the proportions other than A, R, P.

This is an easy proof for the first 3 following statements since each property generates a set of 4 valid valuations (and two of them yield 6 valid valuations). For instance, *sameness* $(T(a, a, b, b)$ implies that valuations $1111, 0000, 0011, 1100$ should be valid and *reflexivity* $(T(a, b, a, b))$ implies that valuations $1111, 0000, 0101, 1010$, which is the truth table of A.

Let us consider the last statement, having the simultaneous satisfaction of the 3 properties leads to a truth table where the 8 valuations $0000, 1111, 1010, 0101,$

20

0110, 1001, 0011, 1100 are valid: then this cannot be the truth table of a logical proportion. $\qquad\square$

It is well known that a valid numerical proportion still holds when we exchange the extreme elements or the mean elements. And we have seen that A and I satisfy both of these permutations. In fact, there are 6 pairwise permutations of the 4 variables appearing in a proportion. So, the behavior of logical proportions w.r.t. these permutations is worth investigating. We denote the permutation of element i and j by $p_{i,j}$: for instance $p_{2,3}$ is the mean permutation while $p_{1,4}$ is the extreme permutation. We can establish the following result:

Proposition 7.
- A is the only proportion to satisfy reflexivity and to be stable for $p_{1,4}$ (or $p_{2,3}$).

- A is the only proportion to satisfy sameness and to be stable for $p_{1,4}$ (or $p_{2,3}$).

- R is the only proportion to satisfy sameness and to be stable for $p_{1,3}$ (or $p_{2,4}$).

- R is the only proportion to satisfy reverse reflexivity and to be stable for $p_{1,3}$ (or $p_{2,4}$).

- P is the only proportion to satisfy reflexivity and to be stable for $p_{1,2}$ (or $p_{3,4}$).

- P is the only proportion to satisfy reverse reflexivity and to be stable for $p_{1,2}$ (or $p_{3,4}$).

- A and I are the only proportions to satisfy symmetry and to be stable for $p_{1,4}$ (or $p_{2,3}$).

- P and I are the only proportions to satisfy symmetry and to be stable for $p_{1,2}$ (or $p_{3,4}$).

- I is the unique logical proportion to satisfy the 6 permutations.

Proof: The proofs are quite similar for the 8 first statements. Let us give an example for the first statement. *reflexivity* means that valuations $0000, 1111, 0011, 1100$ have to be valid. Adding stability for $p_{2,3}$ leads to add 0101 and 1010 as valid valuations. This is the truth table of A.

Let us consider the last statement which is a bit more tricky. It is easy to check that these permutations induce a partition of the set of valuations into 5 classes, each of them being closed for these 6 permutations:

- the class $\{0000\}$ and the class $\{1111\}$

21

- the class $\{0111, 1011, 1101, 1110\}$

- the class $\{1000, 0100, 0010, 0001\}$

- the class $\{0101, 1100, 0011, 1010, 1001, 0110\}$

Taking into account that a logical proportion is true for only 6 valuations (Proposition 1), we only have 3 options:
- a proportion valid for $\{0000\}$, $\{1111\}$ and $\{0111, 1011, 1101, 1110\}$,
- or for $\{0000\}$, $\{1111\}$ and $\{1000, 0100, 0010, 0001\}$,
- or for $\{0101, 1100, 0011, 1010, 1001, 0110\}$.

It appears that the latter class is just the truth table of inverse paralogy. Lemma 3 that we shall prove below allows us to complete the proof. $\qquad\square$

Lemma 3. *A logical proportion cannot satisfies the class of valuation*

$$\{0111, 1011, 1101, 1110\} \ \ or \ the \ class \ \{1000, 0100, 0010, 0001\}.$$

Proof: It is enough to show that this is the case for an equivalence between indicators. So let us consider such an equivalence $l_1 \wedge l_2 \equiv l_3 \wedge l_4$. If this equivalence is valid for $\{0111, 1011\}$, it means that its truth value does not change when we switch the truth value of the 2 first literals from 0 to 1: there are only 2 indicators for a and b satisfying this requirement: $a \wedge b$ and $\overline{a} \wedge \overline{b}$. On top of that, if this equivalence is still valid for $\{1101, 1110\}$, it means that its truth value does not change when we switch the truth value of the 2 last literals from 0 to 1: there are only 2 indicators for c and d satisfying this requirement: $c \wedge d$ and $\overline{c} \wedge \overline{d}$. Then the equivalence $l_1 \wedge l_2 \equiv l_3 \wedge l_4$ is just $a \wedge b \equiv c \wedge d$, $a \wedge b \equiv \overline{c} \wedge \overline{d}$, $a \wedge b \equiv \overline{c} \wedge \overline{d}$ or $\overline{a} \wedge \overline{b} \equiv \overline{c} \wedge \overline{d}$. None of these equivalences satisfies the whole class $\{0111, 1011, 1101, 1110\}$. The same reasoning applies for the other class. $\qquad\square$

We summarize the results of this subsection by a pair of properties characterizing a subset of homogeneous proportions, in Table 6 and Table 7. An empty cell means that the corresponding properties do not characterize any subset of homogeneous proportion. For instance, the diagonal cells are all empty because an homogeneous proportion cannot be characterized with only one property.

To conclude this section, we establish a result which shows how singular I is among the set of homogeneous proportions.

Table 6: Characteristic properties of A, R, P, I

	full identity	code indep.	symmetry	sameness	reflexivity	rev. reflexivity
full identity		A, R, P				
code indep.	A, R, P		A, R, P, I	A, R	A, P	R, P
symmetry		A, R, P, I		A, R		R, P
sameness		A, R	A, R		A	R
reflexivity		A, P		A		P
rev. reflexivity		R, P	R, P	R	P	

Table 7: Characteristic properties of A, R, P, I w.r.t. permutations

	$p_{1,2}$	$p_{1,3}$	$p_{1,4}$	$p_{2,3}$	$p_{2,4}$	$p_{3,4}$
sameness		R	A	A	R	
reflexivity	P		A	A		P
rev. reflexivity	P	R			R	P
symmetry	P, I		A, I	A, I		P, I

Proposition 8.

- *A logical proportion satisfying 2 properties among semi-mirroring, negation-compatibility and exchange-mirroring satisfies the remaining one, and is unique. This is the inverse paralogy I.*

- *A logical proportion stable for 4 permutations is stable for the 2 remaining ones and is unique. This is the inverse paralogy I.*

Proof: Considering the first statement, let us choose for instance *semi-mirroring* and *negation-compatibility*. First of all, we can observe that, for a proportion T to satisfy *semi-mirroring*, means the 4 valuations 1010, 1001, 0110, 0101 are valid. For *negation-compatibility* to be satisfied, the 4 valuations 1100, 0011, 1001, 0110 should be valid. Then the truth table of a proportion satisfying both properties should contains all these valuations i.e. 1010, 1001, 0110, 0101, 1100, 0011. Thanks to Proposition 1, this is the truth table of inverse paralogy I. A similar reasoning applies for the other cases. Regarding the second statement, let us consider a proportion stable for 4 pairwise permutations: since such pairwise permutations generate the full group of permutations of 4 elements, it means this proportion is stable for any permutations. We can consider 2 cases:

- either such a proportion is valid for a valuation having an even number of 0 and other than 0000 and 1111. We can consider this is 0110 for instance. The stability leads to have $0011, 0110, 0101, 1001, 1010$ valid as well: this is the truth table for i.

23

- or such a proportion does not have a valid valuation with an even number of 0 other than 0000 and 1111. It means there is a valid valuation with an odd number of 0 like 1000. In that case, the stability w.r.t. the permutations leads to have $1000, 0100, 0010, 0001$ as valid valuations, which is not possible thanks to Lemma 3. □

3.4 Equation solving

The idea of proportion is closely related to the idea of extrapolation, i.e. to guess/-compute a new value on the ground of existing values. In the case of geometrical proportions, this leads to the well known "rule of three" where, knowing that $\frac{a}{b} = \frac{c}{x}$ holds, allows us to compute the value of x from a, b, c. In the Boolean setting, if for some reason it is believed or known that a logical proportion holds between 4 binary items, 3 of them being known, then one may try to infer the value of the 4th one, at least when this extrapolation leads to a unique value. For a proportion T, there are exactly 6 valuations v such that:

$$v(T(a, b, c, d)) = 1$$

In our context, the problem can be stated as follows. Given a logical proportion T and a valuation v such that $v(a), v(b), v(c)$ are known, does it exist a Boolean value x such that $v(T(a, b, c, d)) = 1$ when $v(d) = x$, and in that case, is this value unique?

We will refer to this problem as "the equation solving problem", and for the sake of simplicity, a propositional variable a is denoted as its truth value $v(a)$, and we use the equational notation $T(a, b, c, x) = 1$, where $x \in \{0, 1\}$ is unknown. First of all, it is easy to see that there are always cases where the equation has no solution. Indeed, the triple a, b, c may take $2^3 = 8$ values, while any proportion T is true only for 6 distinct valuations, leaving at least 2 cases with no solution. For instance, when we deal with analogy A, the equations $A(1, 0, 0, x)$ and $A(0, 1, 1, x)$ have no solution. We have the following results:

Proposition 9.
The analogical equation $A(a, b, c, x)$ is solvable iff $(a \equiv b) \vee (a \equiv c)$ holds. In that case, the unique solution is $x = a \equiv (b \equiv c)$.
The reverse analogical equation $R(a, b, c, x)$ is solvable iff $(b \equiv a) \vee (b \equiv c)$ holds. In that case, the unique solution is $x = b \equiv (a \equiv c)$.
The paralogical equation $P(a, b, c, x)$ is solvable iff $(c \equiv b) \vee (c \equiv a)$ holds.
In each of the three above cases, when it exists, the unique solution is given by $x = c \equiv (a \equiv b)$, i.e. $x = a \equiv b \equiv c$.
The inverse paralogical equation $I(a, b, c, x)$ is solvable iff $(a \not\equiv b) \vee (b \not\equiv c)$ holds. In that case, the unique solution is $x = c \not\equiv (a \not\equiv b)$.

Proof: By immediate investigation of the truth tables. □

The anthropologist, linguist and computer scientist Sheldon Klein [9, 10] was the first to propose to solve analogical equations of the form $A(a, b, c, x) = 1$, where x is unknown, as $x = c \equiv (a \equiv b)$, without however providing an explicit definition for $A(a, b, c, d)$, nor distinguishing between A, R, and P. As we can see, the first 3 homogeneous proportions A, R, P behave similarly. Still, their conditions of equation solvability differ. Moreover, it can be checked that at least 2 of these proportions are always simultaneously solvable. Besides, when they are solvable, there is a common expression that yields the solution.

3.5 Alternative writings for homogeneous proportions

When sticking to the Boolean setting, we can use standard equivalences to get alternative writings for A, R, P, I. First of all, using the De Morgan's laws and the fact that $p \equiv q$ is equivalent to $\bar{p} \equiv \bar{q}$, we get definitions where the internal $\wedge$ are replaced with $\vee$ as shown in Table 8. It means that, in a Boolean setting, indicators involving $\vee$ are a perfect replacement for indicators using $\wedge$.

Table 8: A, R, P, I definitions with $\vee$ operator

A	R
$(a \vee \bar{b} \equiv c \vee \bar{d}) \wedge (\bar{a} \vee b \equiv \bar{c} \vee d)$	$(a \vee \bar{b} \equiv \bar{c} \vee d) \wedge (a \vee \bar{b} \equiv c \vee \bar{d})$
P	I
$(a \vee b \equiv c \vee d) \wedge (\bar{a} \vee \bar{b} \equiv \bar{c} \vee \bar{d})$	$(a \vee b \equiv \bar{c} \vee \bar{d}) \wedge (\bar{a} \vee \bar{b} \equiv c \vee d)$

A more interesting option is to start from the definition of P with indicators

$$(a \wedge b \equiv c \wedge d) \wedge (\bar{a} \wedge \bar{b} \equiv \bar{c} \wedge \bar{d}) \ (P)$$

and to use again De Morgan's laws to rewrite the second equivalence. This leads to a definition of P without any negation that we denote P^*:

$$(a \wedge b \equiv c \wedge d) \wedge (a \vee b \equiv c \vee d) \ (P^*)$$

Then, considering the link between A and P established in Proposition 4, namely $A(a, b, c, d) \equiv P(a, d, c, b)$, it comes another definition for A, without any negation operator:

$$(a \wedge d \equiv b \wedge c) \wedge (a \vee d \equiv b \vee c) \ (A^*)$$

It is noticeable that this latter new definition exactly corresponds to what the psychologist Jean Piaget [15], called *logical proportion*! However, strangely enough, he has not developed their study nor pointed out their link with analogy.

Thus, since a and d are the extreme variables, b and c the mean variables, the analogical proportion $A(a, b, c, d)$ can be read as "the conjunction (resp. disjunction) of the extremes is equivalent to the conjunction (resp. disjunction) of the means".

Considering the link between A, R, P, I coming from Proposition 4, we can finally get alternative writing denoted A^*, R^*, P^* and I^* that are shown in Table 9.

Table 9: A^*, R^*, P^*, I^* definitions

A^*	R^*
$(a \wedge d \equiv b \wedge c) \wedge (a \vee d \equiv b \vee c)$	$(a \wedge c \equiv b \wedge d) \wedge (a \vee c \equiv b \vee d)$
P^*	I^*
$(a \wedge b \equiv c \wedge d) \wedge (a \vee b \equiv c \vee d)$	$(a \wedge b \equiv \overline{c} \wedge \overline{d}) \wedge (a \vee b \equiv \overline{c} \vee \overline{d})$

Since, in the Boolean setting, the equivalence $T(a, b, c, d) \equiv T^*(a, b, c, d)$ holds (where T denotes any homogeneous proportion among A, R, P, I), one could consider T^* as an alternative writing for T. It is interesting to note that this approach leads to rewrite A, R, P without any negation. We have to be aware that these equivalences, leading to alternative writings, are not necessarily valid outside the Boolean framework.

4   Homogeneous proportions: multiple-valued semantics

Ultimately, logical proportions, and in particular the homogeneous ones, could be used for practical applications where we have to deal with missing information or features whose satisfaction is a matter of degree. To cover such situations, extensions of the Boolean interpretation to multiple-valued logics (3-valued at least) is necessary. A formal way to cope with these situations is to extend the Boolean framework to a multiple-valued one by introducing truth values belonging to $[0, 1]$. We should carefully distinguish between three cases:

- when feature satisfaction is a matter of degree instead of being binary, i.e., the truth value of a given feature may be an *intermediate* value between 0 and 1.

- when a feature does not make sense for a given item, i.e., the feature is *non applicable* to it.

- when *information* about some features *is missing*, i.e., we have no clue about the truth value of some features for some items, and the corresponding truth value is not known, i.e., *unknown*.

At this stage, two questions arise:

1. in a given model, what are the valuations that correspond to a "perfect" proportion of a given type (i.e., having 1 as truth value)? For instance, does $T(a, a, a, a)$ postulate still have to be satisfied by A, R, P, or can we consider models where $A(u, u, u, u) = u$, u being a truth value distinct from 0 and 1?

2. are there valuations that could be regarded as "imperfect" proportions of a given type (i.e., with a truth value distinct from 0 and 1) and in that case, what is their truth value?

We investigate these issues in the following subsections keeping in mind an essential principle: whatever the way we define the truth values, the Boolean model should be the limit case of our models when restricted to Boolean valuations.

4.1 Semantics for gradual features

When the satisfaction of features may be a matter of degree, we have to consider that the truth values belong to a linearly ordered scale $\mathcal{L}$. The simplest case is when $\mathcal{L} = \{0, \alpha, 1\}$, with the ordering $0 < \alpha < 1$, which can be generalized into a finite chain $\mathcal{L} = \{\alpha_0 = 0, \alpha_1, \cdots, \alpha_n = 1\}$ of ordered grades $0 < \alpha_1 < \cdots < 1$, or to an infinite chain using the real interval $[0, 1]$. A proposal for extending A in such cases has been advocated in [18]. It takes its source in the initial definition

$$A(a, b, c, d) = (a \wedge \bar{b} \equiv c \wedge \bar{d}) \wedge (\bar{a} \wedge b \equiv \bar{c} \wedge d),$$

where now

- i) the central $\wedge$ is taken as equal to min;

- ii) $s \equiv t$ is taken as $\min(s \rightarrow_L t, t \rightarrow_L s)$ where $\rightarrow_L$ is Łukasiewicz implication, defined by $s \rightarrow_L t = \min(1, 1 - s + t)$, for $\mathcal{L} = [0, 1]$ (in the discrete cases, we take $\alpha = 1/2$ and $\alpha_i = i/n$), and thus $s \equiv t = 1 - |s - t|$; note that $s \equiv t = (1 - s) \equiv (1 - t)$;

- iii) $s \wedge \bar{t} = \max(0, s - t) = 1 - (s \rightarrow_L t)$, i.e., $\wedge^-$ is understood as expressing a bounded difference. Note that this choice preserves $A(a, b, c, d) = A(\bar{a}, \bar{b}, \bar{c}, \bar{d})$ for the involutive negation $\bar{x} = 1 - x$.

The resulting expression for $A(a,b,c,d)$ is given in Table 10. Then, we understand the truth value of $A(a,b,c,d)$ as the extent to which the truth values a,b,c,d makes an analogical proportion. For instance, in such a graded model, the truth value of $A(0.9,1,1,1) = 0.9$, which fits the intuition. It can be checked that the semantics of $A(a,b,c,d)$ thus defined in the graded case, reduces to the previous definition when restricted to the Boolean case.

It is interesting to study in what cases $A(a,b,c,d) = 1$ (and in what cases $A(a,b,c,d) = 0$). Then it is clear that $A(a,b,c,d) = 1$ when $a - b = c - d$. When $a,b,c,d \in \{0, \alpha = 1/2, 1\}$, it yields the 19 following patterns 1111; 0000; $\alpha\alpha\alpha\alpha$; 1010; 0101; $1\alpha1\alpha$; $\alpha1\alpha1$; $0\alpha0\alpha$; $\alpha0\alpha0$; 1100; 0011; $11\alpha\alpha$; $\alpha\alpha11$; $\alpha\alpha00$; $00\alpha\alpha$; $1\alpha\alpha0$; $0\alpha\alpha1$; $\alpha10\alpha$; $\alpha01\alpha$.

This means that $A(a,b,c,d) = 1$ when the change from a to b has the same direction and the same intensity as the change from c to d. However, the last 4 patterns show that there is no need to have $a = b$ and $a = c$ while these conditions hold for the 15 first patterns, which are all of the form $xyxy$, $xxyy$, or $xxxx$. In contrast, note that the last 4 patterns exhibit 3 distinct values.

Table 10: Graded definitions for A, R, P^*

$A(a,b,c,d) =$
$1 - \| (a-b) - (c-d) \|$ if $a \geq b$ and $c \geq d$, or $a \leq b$ and $c \leq d$
$1 - max(\|a-b\|,\|c-d\|)$ if $a \leq b$ and $c \geq d$, or $a \geq b$ and $c \leq d$
$R(a,b,c,d) = A(a,b,d,c)$
$P^*(a,b,c,d) =$
$min(1 - \|max(a,b) - max(c,d)\|, 1 - \|min(a,b) - min(c,d)\|)$

$A(a,b,c,d) = 0$ when $a - b = 1$ and $c \leq d$, or $b - a = 1$ and $d \leq c$, or $a \leq b$ and $c - d = 1$, or $b \leq a$ and $d - c = 1$. It means the 22 following patterns in the 3-valued case: 1110; 1101; 1011; 0111; 0001; 0010; 0100; 1000; 1001; 0110; $10\alpha\alpha$; $01\alpha\alpha$; $\alpha\alpha10$; $\alpha\alpha01$; 100α; 011α; $10\alpha1$; $\alpha001$; $0\alpha10$; $1\alpha01$; $01\alpha0$; $\alpha110$. Thus, $A(a,b,c,d) = 0$ when one change inside the pairs (a,b) and (c,d) is maximal, while the other pair shows no change or a change in the opposite direction.

Using $\mathcal{L} = \{0, \alpha, 1\}$, $A(a,b,c,d) = \alpha$ for 81 - 19 - 22 = 40 distinct patterns.

In [18], the graded extension of $R(a,b,c,d)$ is defined by permuting c and d in the definition of A, according to Proposition 4. But the extension of the paralogy is no longer obtained by permuting b and d in the definition of A (as Proposition 4 would suggest). In fact, the paralogical proportion is defined directly from P^*

(thus changing $\bar{a} \wedge \bar{b} \equiv \bar{c} \wedge \bar{d}$ into $a \vee b \equiv c \vee d$), and taking $\wedge = min$, $\vee = max$, and $s \equiv t = 1 - |s - t|$, we obtain the definition in Table 10. If we now exchange b and d (using Proposition 4 again) in this definition, we get the graded version of A^* (which is no longer equivalent to A), namely

$$A^*(a, b, c, d) = min(1 - |max(a, d) - max(b, c)|, 1 - |min(a, d) - min(b, c)|)$$

This is the direct counterpart of the definition without negation of the analogical proportion in the Boolean case. This alternative extension still preserves $A^*(a, b, c, d) = A^*(\bar{a}, \bar{b}, \bar{c}, \bar{d})$ for the involutive negation $\bar{x} = 1 - x$. It can be checked that $A^*(a, b, c, d) = 1$ only for the 15 patterns with at most two distinct values (for which $A(a, b, c, d) = 1$), while $A^*(a, b, c, d) = \alpha$ for the 4 other patterns for which $A(a, b, c, d) = 1$, namely for $1\alpha\alpha 0$; $0\alpha\alpha 1$; $\alpha 10\alpha$; $\alpha 01\alpha$. Besides, $A^*(a, b, c, d) = 0$ for only 18 among the 22 patterns that make $A(a, b, c, d) = 0$. The 4 patterns for which $A^*(a, b, c, d) = \alpha$ (instead of 0) are $10\alpha\alpha$; $01\alpha\alpha$; $\alpha\alpha 10$; $\alpha\alpha 01$.

Using $\mathcal{L} = \{0, \alpha, 1\}$, $A^*(a, b, c, d) = \alpha$ for 81 - 15 - 18 = 48 distinct patterns.

Thus, it appears that $A^*(a, b, c, d)$ does not acknowledge as perfect the analogical proportion patterns where the amount of change between a and b is the same as between c and d and has the same direction, but where this change applies in different areas of the truth scale. Still, $A^*(a, b, c, d)$ remains half-true in these cases, for $\mathcal{L} = \{0, \alpha, 1\}$. When $\mathcal{L} = [0, 1]$, it can be checked that $A^*(a, b, c, d) \geq 1/2$ when $a - b = c - d$; in particular, $\forall a, b, A^*(a, b, a, b) = 1$, which corresponds to the case where $a = c$ and $b = d$. In the same spirit, if $\mathcal{L} = \{0, \alpha, 1\}$ as well as for $\mathcal{L} = [0, 1]$, $A^*(a, b, c, d) = 0$ when a change inside the pairs (a,b) and (c,d) is maximal, while the other pair shows a change in the opposite direction starting from 0 or 1. However, $A^*(1, 0, c, c) = min(c, 1-c)$ and A^* takes the same value for the 7 other permutations of $(1, 0, c, c)$ obtained by applying symmetry and/or central permutation.

As can be seen in Table 11, A^* and A also coincide on some patterns having intermediary truth values, but diverge on others. Generally speaking, A^* is smoother than A in the sense that more patterns have intermediary truth values with A^* than with A.

Both A and A^* continue to satisfy the *symmetry property* (as P, R, and P^*, R^* with $R^*(a, b, c, d) = A^*(a, b, d, c) = P^*(a, c, d, b))$. However, only A^* still enjoys the *means permutation* and the *extremes permutation* properties. *This is no longer the case with A*, as shown by the following counter-example.

$A(0.8, 0.6, 1, 0.3) = 1 - |(0.8 - 0.6) - (1 - 0.3)| = 1 - |0.2 - 0.7| = 0.5$ since $0.8 \geq 0.6$ and $1 \geq 0.3$, and $A(0.8, 1, 0.6, 0.3) = 1 - max(|0.8 - 1|, |0.6 - 0.3|) = 1 - max(0.2, 0.3) = 0.7$ since $0.8 \leq 1$ and $0.6 \geq 0.3$.

Table 11: The two graded definitions of the analogical proportion in $[0, 1]$

A	A^*
$A(1,1,u,v) = 1 - \|u-v\|$	$A^*(1,1,u,v) = 1 - \|u-v\|$
$A(1,0,u,v) = u - v$ if $u \geq v$	$A^*(1,0,u,v) = \min(u, 1-v)$
$\quad = 0$ if $u \leq v$	
$A(0,1,u,v) = v - u$ if $u \leq v$	$A^*(0,1,u,v) = min(v, 1-u)$
$\quad = 0$ if $u \geq v$	
$A(0,0,u,v) = A(1,1,u,v)$	$A^*(0,0,u,v) = A^*(1,1,u,v)$

But, as already mentioned, *both A and A^** continue to satisfy the *code independency* property with respect to $\bar{a} = 1 - a$. Some more Boolean properties that remain valid in the multiple-valued case are summarized in Table 12.

Table 12: Graded properties of A, A^*, R, P

Property name	Formal definition	Proportion
full identity	$T(a,a,a,a)$	A^*, A, R, P
reflexivity	$T(a,b,a,b)$	A^*, A, P
reverse reflexivity	$T(a,b,b,a)$	R,P
sameness	$T(a,a,b,b)$	A^*, A, R
symmetry	$T(a,b,c,d) \to T(c,d,a,b)$	A^*, A, R, P
permutation of means	$T(a,b,c,d) \to T(a,c,b,d)$	A^*
permutation of extremes	$T(a,b,c,d) \to T(d,b,c,a)$	A^*
all permutations	$\forall i,j, T(a,b,c,d) \to T(p_{i,j}(a,b,c,d))$	none
semi-mirroring	$T(a,b,\bar{a},\bar{b})$	R
exchange mirroring	$T(a,b,\bar{b},\bar{a})$	A
negation compatib.	$T(a,\bar{a},b,\bar{b})$	none

4.2   Dealing with non-applicable features

The abbreviation 'n/a' (for *non applicable*) is currently used in data tables when an attribute does not apply, when a feature does not make sense for a particular item. However, the extensive use of 'n/a' may be often ambiguous when it also appears in the same tables when information is *non available* for some attribute values of some items. Indeed one has to carefully distinguish the case where the feature does apply to the item, but it is not known if the feature is true or is false for the item, from the

30

case where the feature is neither true nor false for the item since the feature does not apply to it. The case of unknown truth values is discussed in the next section, while we now address the problem of dealing with genuinely non applicable features.

The idea of introducing a third truth value for 'non applicable' (na for short in the following) in the context of analogy can be already found in the pioneering work of Sheldon Klein [9, 10], which we already mentioned in the equation solving subsection 3.4. However, his handling of na is based on $(na \equiv na) = na$, which suggests that the evaluation of an analogical proportion where na appears may receive the truth value na, which is more in the spirit of understanding na as 'not available', or 'unknown'.

Indeed, although a property may be 'true', 'false', or 'non applicable' for an item, it seems natural to expect that $A(a, b, c, d)$ can only be 'true' or 'false', since $1na1na$ looks intuitively satisfactory as an analogical proportion, while $1na00$ is not. More precisely, in the context of non applicable properties, we have only 3 valuation patterns that should make an analogical proportion true: $xxxx$, $xyxy$, and $xxyy$, where $x, y \in \{0, 1, na\}$. Any other option should make it false, since $\{0, 1, na\}$ play the same role. This leads to acknowledge as true the 15 following valuations:

- 1111; 0000; $nananana$ corresponding to $xxxx$;
- 1010; 0101; $1na1na$; $na1na1$; $0na0na$; $na0na0$ corresponding to $xyxy$ with $x \neq y$;
- 1100; 0011; $11nana$; $nana11$; $nana00$; $00nana$ corresponding to $xxyy$ with $x \neq y$.

All the remaining valuations lead to false.

In other words, we are in a situation somewhat similar to the one encountered in the previous section in the case of a unique intermediary truth-value α between true and false, meaning 'half-true' (or equivalently 'half-false'), when we refuse the four valuations $1\alpha\alpha0$, $0\alpha\alpha1$, $\alpha01\alpha$ and $\alpha10\alpha$ as being true, *except that* now no valuation leads to the third truth value. It is possible to find logical definitions of the analogical proportion having the expected behavior for the truth values $\{0, 1, na\}$. A solution to get the exact truth table is:
- to order $\{0, 1, na\}$ as the chain $1 > na > 0$,
- to use Kleene conjunction and disjunction, see, e.g., [2], respectively defined by the minimum and the maximum according to the above ordering,
- to use the strong Kleene equivalence $\equiv$, where $x \equiv y = 1$ if and only if $x = y$, and $x \equiv y = 0$ otherwise,
- to define analogical proportion with A^* instead of A, namely

$$A^*(a, b, c, d) = (a \wedge d \equiv b \wedge c) \wedge (a \vee d \equiv b \vee c).$$

A counterpart to $A(a, b, c, d) = (a \setminus b \equiv c \setminus d) \wedge (b \setminus a \equiv d \setminus c)$ where $\setminus$ here

31

denotes the Boolean logical connective corresponding to set difference, can also be found. However, since we do not want to have $1nana0$ true, the difference between 1 and na and the difference between na and 0 should not be the same, neither the same as between 1 and 0, nor between 1 and 1 for sure. Thus we need 4 distinct values for the difference. This is impossible with 3 truth values! This contrasts with the Boolean case where there are only two possible difference values needed. The solution is then to use 2 connectives for differences:

$x \setminus_1 y = 1$ if $x=1$ and $y=0$; $x \setminus_1 y = na$ if $x=1$ and $y=na$; $x \setminus_1 y = 0$ otherwise;

$x \setminus_2 y = 1$ if $x=1$ and $y=0$; $x \setminus_2 y = na$ if $x=na$ and $y=0$; $x \setminus_2 y = 0$ otherwise.

Then the definition of $A(a,b,c,d)$ becomes

$$(a \setminus_1 b \equiv c \setminus_1 d) \wedge (b \setminus_2 a \equiv d \setminus_2 c) \wedge (a \setminus_2 b \equiv c \setminus_2 d) \wedge (b \setminus_1 a \equiv d \setminus_1 c)$$

where $x \equiv y = 1$ iff $x = y$; $x \equiv y = 0$ otherwise; and $\wedge$ is any conjunction connective that coincides with classical conjunction on $\{0,1\}$. This definition yields 1 for the 15 expected patterns and is 0 otherwise for the $81 - 15 = 66$ remaining patterns.

It is even possible to find an expression for $A(a,b,c,d)$ where $\setminus_1$ and $\setminus_2$ are expressed in terms of a conjunction (denoted $\wedge^*$) and two distinct negation operators $(\cdot)^1$ and $(\cdot)^2$, i.e., where $x \setminus_1 y$ is replaced by $x \wedge^* \bar{y}^1$ and $x \setminus_2 y$ is replaced by $x \wedge^* \bar{y}^2$. We obtain a definition for $A(a,b,c,d)$ under the form

$$(a \wedge^* \bar{b}^1 \equiv c \wedge^* \bar{d}^1) \wedge^* (b \wedge^* \bar{a}^2 \equiv d \wedge^* \bar{c}^2) \wedge^* (a \wedge^* \bar{b}^2 \equiv c \wedge^* \bar{d}^2) \wedge^* (b \wedge^* \bar{a}^1 \equiv d \wedge^* \bar{c}^1)$$

where the two negations are Post-like negations defined through a circular ordering of the three truth-values, where the negation of a value is the successor value in the ordering, namely $\bar{0}^1 = na; \overline{na}^1 = 1; \bar{1}^1 = 0$ and $\bar{0}^2 = 1; \overline{na}^2 = 0; \bar{1}^2 = na$. This acknowledges the fact that in some sense these three truth-values play similar roles. The non-standard three-valued conjunction $\wedge^*$, which is defined by

$x \wedge^* y = 1$ if $x = 1$ and $y = 1$

$x \wedge^* y = u$ if $x = na$ and $y = na$

$x \wedge^* y = 0$ otherwise

also agrees with this view, while coinciding with classical conjunction in the binary case.

As in the previous section, we summarize in Table 13 the properties of the Boolean case that remain valid in this 3-valued model where na, standing for non applicable, is the third truth value.

4.3 Dealing with unknown features

In this section, we briefly consider a situation that is quite different from the ones studied in the two previous sections. We assume now that the features used for

Table 13: Properties of A, R, P with truth value na (as non applicable)

Property name	Formal definition	Proportion
full identity	$T(a, a, a, a)$	A,R,P
reflexivity	$T(a, b, a, b)$	A,P
reverse reflexivity	$T(a, b, b, a)$	R,P
sameness	$T(a, a, b, b)$	A,R
symmetry	$T(a, b, c, d) \to T(c, d, a, b)$	A,R,P
permutation of means	$T(a, b, c, d) \to T(a, c, b, d)$	A
permutation of extremes	$T(a, b, c, d) \to T(d, b, c, a)$	A
all permutations	$\forall i, j, T(a, b, c, d) \to T(p_{i,j}(a, b, c, d))$	none

describing situations are all binary (i.e., they can be only true or false), but their truth value may be unknown.

Thus, the possible states of information regarding a Boolean variable x pertaining to a given feature may be represented by one of the 3 truth value subsets $\{0\}, \{1\}$ or $\{0, 1\}$, corresponding respectively to the case where the truth value of x is false, true or unknown. We denote this state of information by $\tilde{x}$, which is a subset of $\{0, 1\}$. The evaluation of a logical proportion $T(a, b, c, d)$ then amounts to compute the state of information denoted $\mathcal{T}(\tilde{a}, \tilde{b}, \tilde{c}, \tilde{d})$ about its truth value, knowing $\tilde{a}, \tilde{b}, \tilde{c}, \tilde{d}$. It is given by the standard set extension:

$$\mathcal{T}(\tilde{a}, \tilde{b}, \tilde{c}, \tilde{d}) = \{v(T(a, b, c, d)) \mid v(a) \in \tilde{a}, v(b) \in \tilde{b}, v(c) \in \tilde{c}, v(d) \in \tilde{d}\}$$

where v denotes a Boolean valuation.

From now on, we focus on analogical proportion A only, but R, P and I could be handled in a similar manner. For instance, let us take the example $A(a, b, c, d)$ where $\tilde{a} = \{1\}, \tilde{b} = \{0\}, \tilde{c} = \tilde{d} = \{0, 1\}$. Applying the previous formula leads to

$$\mathcal{A}(\tilde{a}, \tilde{b}, \tilde{c}, \tilde{d}) = \{0, 1\}$$

since the truth value of $A(a, b, c, d)$ may be 0 for the valuations 1001, 1000, 1011, and 1 for 1010.

Let us now consider the following expression $A(a, b, a, b)$ when $\tilde{a} = \tilde{b} = \{0, 1\}$. A similar computation leads to

$$\mathcal{A}(\tilde{a}, \tilde{b}, \tilde{a}, \tilde{b}) = \{1\}$$

since the truth value of $A(a, b, a, b)$ is 1 for any of the valuations 1010, 1111, 0101, or 0000. Similarly, the truth value of $A(a, a, a, a)$ is 1, even when $\tilde{a} = \{0, 1\}$.

33

But, the set of possible truth values for $A(a, b, c, d)$ is $\{0, 1\}$ when $\tilde{a} = \{0, 1\}, \tilde{b} = \{0, 1\}, \tilde{c} = \{0, 1\}, \tilde{d} = \{0, 1\}$. It should be clear that this does not mean that the Boolean variables a, b, c, d are equal; we just have the same state of information for all of them. This expresses that the full identity property does not hold any longer at the information level for analogical proportion. And this illustrates the fact that the logic of uncertainty is no longer truth functional, since the state of information about the truth value of $A(a, b, c, d)$ does not only depend on the state of information about the truth values of $a, b, c,$ and d, but is also constrained by the existence of possible logical dependencies between these variables.

Nevertheless, some key properties of homogeneous proportions remain valid at the information level such as symmetry, or central and extreme permutations. Indeed it can be checked that, for instance, for symmetry:

$$\mathcal{A}(\tilde{a}, \tilde{b}, \tilde{c}, \tilde{d}) = \mathcal{A}(\tilde{c}, \tilde{d}, \tilde{a}, \tilde{b})$$

Using the set extension evaluation of logical proportions in presence of incomplete information, we can compute the set of possible truth values of the analogical proportion for the different 4-tuples of states of information. We now denote by u the state $\{0, 1\}$, and respectively by 0 and 1, the states of information $\{0\}$ and $\{1\}$. A 4-tuple of states of information will be called *information pattern*, or pattern for short, and denoted by a 4-tuple of elements of $\{0, 1, u\}$ without blank space. For instance, $01u1$ is such a pattern and should be understood as the 4-tuple of states of information $(\{0\}, \{1\}, \{0, 1\}, \{1\})$.

Then, the 6 patterns $0000, 1111, 0011, 1100, 1010, 0101$ that makes A true in the Boolean case, and where u does not appear, are the only ones that are still true with the above view (for which we get the singleton $\{1\}$ as information state for $A(a, b, c, d)$). As soon as at least one state of information is u in the pattern, the state of information for $A(a, b, c, d)$ is u or 0. Indeed, for instance, $01u0$ leads to 0 since whatever the truth value of the 3rd variable, the analogical proportion does not hold. Thus, despite the lack of knowledge regarding the 3rd variable, we know the exact truth value of the proportion in this case, namely it is false. It appears that there are 18 patterns that lead to 0. They are the 10 patterns of the Boolean case and the 8 following ones: $01u0, 0u10, u001, 100u, 10u1, 1u01, u110, 011u$. Thus, in the $81 - 6 - 18 = 57$ remaining cases, the state of information for $A(a, b, c, d)$ is u.

It can be checked that these results can be retrieved both with the initial definition of A or with A^* where complete ignorance u is handled with $\bar{\ }, \wedge, \vee$ as the strong Kleene connectives (see [2]) and $\equiv$ as Bochvar connective, where u is an absorbing element. The corresponding truth tables are recalled in Table 14. This provides a way to extend the definition of the analogical proportion in case of lack

Table 14: Truth tables for u as lack of knowledge

	¯	∧	0	1	u	∨	0	1	u	≡	0	1	u
0	1	0	0	0	0	0	0	1	u	0	1	0	u
1	0	1	0	1	u	1	1	1	1	1	0	1	u
u	u	u	0	u	u	u	u	1	u	u	u	u	u

of knowledge when no dependencies between the variables exist. As in the Boolean case, the definitions A (resp. R, P, I) and A^* (resp. R^*, P^*, I^*) are equivalent.

Nevertheless, this truth-functional calculus provides only a description of the evaluation of the patterns at the information level. Namely, it enables us to retrieve the tri-partition of the patterns in respectively 6, 18 and 57 patterns leading respectively to 1, 0 and u, but it does not account for the full calculus of the extended definition of logical proportions in presence of incomplete information, when dependencies take place between variables, for instance it can be checked that $A(a, b, a, b)$ and $A^*(a, b, a, b)$ when a and b are unknown does not yield 1 as expected, but u (this is just due to the fact that constraints $a = c$ and $b = d$ are ignored).

5 Heterogeneous proportions

As highlighted in the introduction, there are 4 other proportions that satisfy code independency, and as such stand out of the 120 logical proportions, namely the heterogeneous proportions, whose truth tables are given in Table 15.

Table 15: H_a, H_b, H_c, H_d - Boolean truth tables

H_a					**H_b**					**H_c**					**H_d**			
1	1	1	0		1	1	1	0		1	1	1	0		1	1	0	1
0	0	0	1		0	0	0	1		0	0	0	1		0	0	1	0
1	1	0	1		1	1	0	1		1	0	1	1		1	0	1	1
0	0	1	0		0	0	1	0		0	1	0	0		0	1	0	0
1	0	1	1		0	1	1	1		0	1	1	1		0	1	1	1
0	1	0	0		1	0	0	0		1	0	0	0		1	0	0	0

It is stunning to note that these truth tables exactly involve the 8 missing tuples of the homogeneous tables, i.e., those ones having an odd number of 0 and 1. It should not come as a surprise that they satisfy the same association properties as the homogeneous ones: for instance, any combination of 2 or 3 heterogeneous proportions is satisfiable, but the conjunction $H_a(a, b, c, d) \wedge H_b(a, b, c, d) \wedge H_c(a, b, c, d) \wedge$

$H_d(a, b, c, d)$ is not satisfiable. This fact contributes to make the heterogeneous proportions the perfect dual of the homogeneous ones.

5.1 Properties

The formal definitions given in Table 2 lead to immediate Boolean equivalences between heterogeneous and homogeneous proportions that we summarize in Table 16.

Table 16: Equivalences between heterogeneous and homogeneous proportions

H$_a$	**H$_b$**
$H_a(a,b,c,d) \equiv I(\overline{a},b,c,d)$	$H_b(a,b,c,d) \equiv I(a,\overline{b},c,d)$
$H_a(a,b,c,d) \equiv P(\overline{a},b,\overline{c},\overline{d})$	$H_b(a,b,c,d) \equiv P(a,\overline{b},\overline{c},\overline{d})$
$H_a(a,b,c,d) \equiv P(a,\overline{b},c,d)$	$H_b(a,b,c,d) \equiv P(\overline{a},b,c,d)$
H$_c$	**H$_d$**
$H_c(a,b,c,d) \equiv I(a,b,\overline{c},d)$	$H_d(a,b,c,d) \equiv I(a,b,c,\overline{d})$
$H_c(a,b,c,d) \equiv P(\overline{a},\overline{b},\overline{c},d)$	$H_d(a,b,c,d) \equiv P(\overline{a},\overline{b},c,\overline{d})$
$H_c(a,b,c,d) \equiv P(a,b,c,\overline{d})$	$H_d(a,b,c,d) \equiv P(a,b,\overline{c},d)$

Obviously, the heterogeneous proportions are strongly linked together: for instance, using the symmetry of I,

$$H_a(a,b,c,d) \equiv I(\overline{a},b,c,d) \equiv I(c,d,\overline{a},b) \equiv H_c(c,d,a,b).$$

We may consider two different ways for generating these proportions:

- A *semantic viewpoint*: The *full identity* postulate $T(a,a,a,a)$ asserts that proportion T holds between identical values. Negating one variable position only generates an intruder, as in $T(\overline{a},a,a,a)$, $T(a,\overline{a},a,a)$, $T(a,a,\overline{a},a)$ and $T(a,a,a,\overline{a})$, and leads to new postulates respectively denoted T_a, T_b, T_c and T_d. We call the negated position the *intruder position*: for instance, T_a expresses the fact that the first position is an intruder. For a proportion, to satisfy the property T_a means that *the first variable may be an intruder*. Since each postulate T_a, T_b, T_c and T_d is validated by 2 distinct valuations, it is clear that 3 of them are enough to define a logical proportion having exactly 6 valid tuples. There is no proportion satisfying all these postulates since it leads to 8 valid tuples, which excludes any logical proportion. It can be easily checked that H_a satisfies T_b, T_c, T_d and does not satisfy T_a: then H_a is uniquely characterized by the conjunction of properties $T_b \wedge T_c \wedge T_d$. We can interpret $H_a(a,b,c,d)$ as the following assertion: *the first position is not an intruder and there is an intruder among the remaining positions.* As a consequence, $H_a(a, b, c, d)$ does not hold when there is no intruder (i.e., when there is an even number of 0), or when a is the intruder. The same reasoning applies to H_b, H_c, H_d.

- A *syntactic viewpoint*: Here we start from the definition of the inverse paralogy I: $(a \wedge b \equiv \overline{c} \wedge \overline{d}) \wedge (\overline{a} \wedge \overline{b} \equiv c \wedge d)$. To get the definition of an heterogeneous proportion satisfying postulates where the intruder is in position 4, 2 or 1 for instance, we add a negation on the 3rd variable in both equivalences defining I. Here we get H_c as:

$$(a \wedge b \equiv c \wedge \overline{d}) \wedge (\overline{a} \wedge \overline{b} \equiv \overline{c} \wedge d)$$

This process, allowing us to generate the 4 heterogeneous proportions, shows that, in some sense, they are "atomic perturbations" of I: for this reason and since they are heterogeneous proportions, they have been respectively denoted H_a, H_b, H_c and H_d where the subscript corresponds to:

- the postulate *which is not satisfied* by the corresponding proportion or, equivalently,

- the negated variable in the equivalence with I.

For instance $H_a(a, b, c, d) \equiv I(\overline{a}, b, c, d)$, H_a satisfies T_b, T_c, T_d and does not satisfy T_a. This leads to another way to interpret $H_a(a, b, c, d)$. Since $H_a(a, b, c, d) \equiv I(\overline{a}, b, c, d)$, when $H_a(a, b, c, d) = 1$, a is not the intruder, i.e., $\overline{a}$ is the value of the intruder. The different possible cases are as follows:

- $\overline{a}bcd = 1100$ or 0011 and the intruder is b,

- or $\overline{a}bcd = 0101$, 0110, 1010 or 1001 and the intruder is c or d.

In other words, there is an intruder in (a, b, c, d), which is not a, iff the properties common to $\overline{a}$ and b (positively or negatively) are not those common to c and d, and conversely.

As in the case of homogeneous proportions, the semantic properties of heterogeneous proportions are easily derived from their truth tables, which we summarize in Table 17. It is clear on their truth tables, that none of the heterogeneous proportions satisfy neither symmetry nor transitivity. From a practical viewpoint, these proportions are closely related with the idea of spotting the odd one out (the intruder), or if we prefer of picking the one that doesn't fit among 4 items. This will be further discussed in Section 6, but we first consider the extension of heterogeneous proportions to the case of graded properties with intermediate truth values.

Table 17: Properties of heterogeneous proportions

Property name	Formal definition	Proportion
full identity	$T(a, a, a, a)$	none
reflexivity	$T(a, b, a, b)$	none
reverse reflexivity	$T(a, b, b, a)$	none
sameness	$T(a, a, b, b)$	none
symmetry	$T(a, b, c, d) \rightarrow T(c, d, a, b)$	none
means permut.	$T(a, b, c, d) \rightarrow T(a, c, b, d)$	H_a, H_d
extremes permut.	$T(a, b, c, d) \rightarrow T(d, b, c, a)$	H_c, H_b
all permutations	$\forall i, j, T(a, b, c, d) \rightarrow T(p_{i,j}(a, b, c, d))$	none
transitivity	$T(a, b, c, d) \wedge T(c, d, e, f) \rightarrow T(a, b, e, f)$	none
Ta	$T(\overline{a}, a, a, a)$	H_b, H_c, H_d
Tb	$T(a, \overline{a}, a, a)$	H_a, H_c, H_d
Tc	$T(a, a, \overline{a}, a)$	H_a, H_b, H_d
Td	$T(a, a, a, \overline{a})$	H_a, H_b, H_c

5.2   Multiple-valued semantics

We extend here what has been done for homogeneous proportions and their multiple-valued semantics in Section 4.1. Roughly speaking, in the case of H_a, the graded truth value of $H_a(a, b, c, d)$ estimates how *far* we are from having a as an intruder.

Obviously the same questions as for homogeneous proportions arise but with a different interpretation:

1) what are the valuations that correspond to a "perfect" proportion of a given type (i.e., having 1 as truth degree)? For instance, we want the truth value of $H_a(0, u, 0, u)$ to be equal to 1 (as well as the truth value of $H_c(0, u, 0, u)$) because in that context, it is true that $a = 0$ (resp. c) cannot be the intruder, whatever the value of u.

2) are there valuations that could be regarded as *approximate* proportions of a given type (with an intermediate truth degree) and in that case, what is their truth value? For instance, in the valuation $(0.7, 1, 1, 0.9)$, it is likely that a is the intruder just because the other candidate, d, has a value very close to 1, and the closer d is to 1, the more likely a is the intruder: then the truth value of $H_a(0.7, 1, 1, 0.9)$ should be small and related to $1 - d = 0.1$ (since H_a excludes a as intruder).

The most rigorous way to proceed is to start from the definition of multiple-valued paralogy given in Section 4.1. This definition is based on P^*: it leads, for a three valued scale, to 15 valuations fully true, and 18 fully false. The 48 remaining patterns get intermediate truth value given by the following general formula

$$P^*(a,b,c,d) = min(1 - |max(a,b) - max(c,d)|, 1 - |min(a,b) - min(c,d)|)$$

which, thanks to the symmetry of P^* and stability w.r.t. the permutation of its two first variables, has the following behavior:

general case	case $u = v$
$P^*(1,1,u,v) = min(u,v)$	$P^*(1,1,u,u) = u$
$P^*(1,0,u,v) = min(max(u,v), 1-min(u,v))$	$P^*(1,0,u,u) = min(u,1-u)$
$P^*(0,0,u,v) = 1 - max(u,v)$	$P^*(0,0,u,u) = 1 - u$

Starting from the equivalences given in Table 16, we get the multi-valued definition for H_a (and similar definitions for H_b, H_c, H_d), still leading to 15 true valuations, 18 false valuations and 48 with intermediate values in case of a three valued scale:

$$H_a(a,b,c,d) = min(1 - |max(a,1-b) - max(c,d)|, 1 - |min(a,1-b) - min(c,d)|)$$

Let us note that $H_a(0,0,u,v) = H_a(1,1,u,v)$ due to the equality $H_a(0,0,u,v) = P(0,1,u,v) = P(1,0,u,v)$. We have:

general case	case $u = v$
$H_a(1,1,u,v) = min(max(u,v), 1-min(u,v))$	$H_a(1,1,u,u) = min(u,1-u)$
$H_a(1,0,u,v) = min(u,v)$	$H_a(1,0,u,u) = u$
$H_a(0,1,u,v) = 1 - max(u,v)$	$H_a(1,0,u,u) = 1 - u$

Let us analyze two examples to highlight the fact that the above definition really fits with the intuition.

- Considering the valuation $100u$, its truth value is:

  - u for P: if u is close to 1, we are close to the fully true paralogical proportion and the truth value is high. In the opposite case, u is close to 0 and we are close to a fully false paralogical proportion 1000.

  - 1-u for H_b, H_c, H_d: if u is close to 1, we are close to the valuation 1001 which is definitely not a valid valuation for H_b, H_c, H_d: so 1-u is a low truth value. But if u is close to 0, we are close to the valuation 1000 which is valid for H_b, H_c, H_d and 1-u is a high truth value.

  - finally 0 for H_a: whatever the value of u, 100u means "an intruder is in first position", when the semantics of H_a is just the opposite.

- Back to the graded valuation valuation 0.7 1 1 0.9 considered above:

- regarding P, the truth value as given by the formula is 0.8, i.e., the valuation is close to be a true paralogy.

- regarding the heterogeneous proportions, we understand that we have 2 candidate intruders namely $a = 0.7$ and $d = 0.9$. But they are not equivalent in terms of intrusion and it is more likely to be a than d. This is consistent with the fact that the truth value of $H_a(0.7, 1, 1, 0.9)$ is 0.1 (very low), but the truth value of $H_d(0.7, 1, 1, 0.9)$ is 0.3 (a bit higher).

- in fact, 0.7 1 1 0.9 does not give a genuine impression that there is an intruder, which is in agreement with the fact that $H_b(0.7, 1, 1, 0.9) = H_c(0.7, 1, 1, 0.9) = 0.4$.

6 Applications

In this section, we provide an overview of the use of logical proportions for various reasoning purposes. Since we have distinguished two remarkable groups of proportions, differing both from a syntactic and a semantic viewpoint, it is not surprising that they can be used for two different styles of applications. On the one hand, the homogeneous proportions allow us to build up a missing item in a given sequence. On the other hand, the heterogeneous proportions are suitable for a dual task which is to pick up the odd one out in a set. Let us start by discussing the use of the homogeneous logical proportions, which is the most developed.

6.1 Using homogeneous proportion for finding missing values

From a general viewpoint, a homogeneous proportion between 4 items a, b, c, d expresses that the elements of the pair (a, b) are similar (or dissimilar) in a way that can be related to the way the elements of the pair (c, d) are similar (or dissimilar). The equation-solving process described above enables us to compute d from the knowledge of a, b, c, when possible. Obviously, in practical cases, the items to be considered cannot be simply described by a single Boolean (or multiple-valued) variable, and a straightforward extension, allowing to cope with more sophisticated representations, can be given for Boolean vectors in $\mathbb{B}^n$, as follows (where T denotes any logical proportion):

$$T(\overrightarrow{a}, \overrightarrow{b}, \overrightarrow{c}, \overrightarrow{d}) \text{ iff } \forall i \in [1, n], \ T(a_i, b_i, c_i, d_i).$$

The solving process of the equation $T(\overrightarrow{a}, \overrightarrow{b}, \overrightarrow{c}, \overrightarrow{x})$ is still effective: instead of getting one Boolean value, we get a Boolean vector, by solving equations componentwise, computing d_i from a_i, b_i, and c_i (provided that the solution exists). This

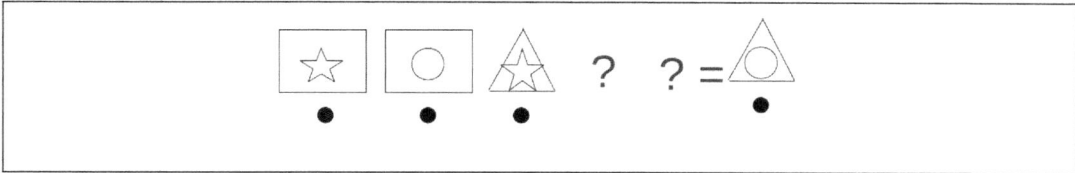

Figure 1: IQ test: Graphical analogy

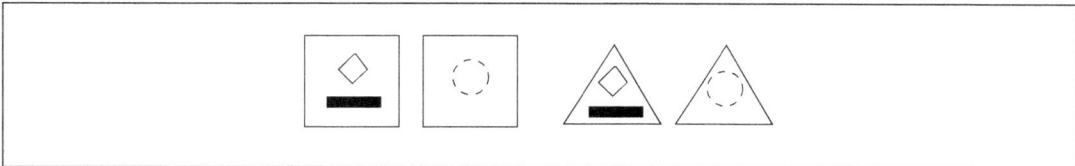

Figure 2: Analogy with a graded feature

can be illustrated on a sequence of 3 pictures to be completed (see Figure 1, as it is often the case in IQ tests . Indeed, a noticeable part of the IQ tests are based on providing incomplete analogical proportions (see, e.g., [6]). Usually, the 3 first items A, B, C are given and the 4th item X has to be chosen among several plausible candidates. In this case, the homogeneous logical proportion method applies straightforwardly. The items A, B, C in the example of Figure 1 can be described respectively by vectors $(1, 0, 1, 0, 1)$, $(1, 0, 0, 1, 1)$, $(0, 1, 1, 0, 1)$, where the vector components refer respectively to the presence (or not) of a square, of a triangle, of a star, of a circle, and of a black point. Assuming that an analogical proportion should hold, by solving componentwise the analogical proportion equations expressing that $A(a_i, b_i, c_i, x_i)$ holds true for $i = 1, 5$, we easily get $X = (0, 1, 0, 1, 1)$, which corresponds to the result exhibited in Figure 1. Note that X is directly computed with this method, rather than chosen among a set of more or less "distant" potential solutions that would be given. In case the analogical equation has no solution for some component, on may try if another homogeneous proportion would fit for all the features. It would not be difficult to build examples of sequences of 4 pictures, where the display of squares, triangles, stars, circles and black dots is different from Figure 1, and where the fourth picture would be obtained via one of the three other homogeneous proportions R, P, or I, rather than via A as in Figure 1.

Moreover, Figure 2 illustrates the idea of having graded features, where here the presence of a circle is a matter of degree (the more densely dotted the circle, the higher the degree α of presence of a circle (in Figure 2 the analogical proportion $A(0, \alpha, 0, \alpha)$ clearly holds for the 'circle' variable).

In the above example, the problem is handled at a rather high conceptual level that requires that triangles, circles and so on be identified in the pictures. However, it has been pointed out [20] that the analogical proportion-based technique can still be applied at the pixel level. Then a black and white picture is represented by the Boolean vector made of its *bitmap* description that acknowledges (or not) the black color of each pixel. This supposes that all the geometric shapes (squares, triangles, stars, circles) use exactly the same pixels in all cases. Then, the proportion-based procedure automatically builds the associated geometric figure (when it exists), without introducing any knowledge about triangle, circle, etc.

Lastly, let us mention that it may be convenient to have extensions of the proportions allowing for the explicit handling of functional symbols, as in, e.g., the analogical proportion $A(x, f(x), y, f(y))$, for handling more sophisticated sequences of pictures (where for instance, elements are reversed from one picture to another), or analogical proportions quizzes like "*abc* is to *abd* as *ijk* is to ?" (where we have to encode that d is the successor of c); see [3].

6.2 Classification and matrix abduction

We now consider variants of the process described in the previous subsection, when it is first checked that an homogeneous proportion holds on a series of n features between 4 items, and on this basis, one extrapolates that the same logical proportion still holds for a $(n+1)$th feature of interest, which is known only for the first 3 items. Solving the logical proportion equation corresponding to this latter feature then enables us to compute a plausible value of this feature for the 4th item. Classification problems are an important instance of this situation where the $(n+1)$th feature refers to the class of the item while the n other features pertain to its description. See [13] for the case of binary features, where very good results are reported on classification benchmarks. The graded version A has been used for handling numerical features in classification problems (also with promising experimental results [23]), while A^* has not been experienced yet. It is still unclear if A^* may be more suitable for classification purposes.

The problem of completing a matrix where some values are missing is quite close to the classification problem, and thus different methods may be thought of in order to deal with this issue. Whatever the technique, the main question is to know if the extra knowledge that we may have about the problem and the available data carry sufficient information for an accurate reconstruction of the missing cells. This is not always the case. We focus here on a particular case, called "matrix abduction problem", using [1]'s terminology. It consists in guessing plausible values for cells having empty information in a matrix where each line corresponds to a situation described according to different binary features (each column corresponds to a particular feature).

Table 18: The screen example

	P	C	I	R	D	S
screen1	0	1	0	1	0	1
screen2	0	0	1	1	0	1
screen3	0	0	0	0	1	?
screen4	1	1	0	0	1	1

Let us consider the screen example used by [1], where computer screens are described by 6 characteristic features: P is for price over £450, C for self collection, I for screen bigger than 24 inch, R for reaction time below 4ms, D for dot size less than 0.275, and S for stereophonic; 1 means "yes" and 0 means "no". We have 3 screens (screen 1, screen 2 and screen 4) whose characteristics are known and screen 3 where the truth value of the attribute S is missing (see Table 18).

To tackle such a common sense problem, a general idea (which may be also found in classification) is that replacing an unknown value by either 1 or 0 should result in the least possible *perturbation* of the matrix. This idea may be implemented in diverse ways. In [1] the idea is to choose the value that least perturbs the pre-existing partial ordering between the column vectors of the matrix. In [25], the idea is rather to respect betweenness and parallelism relations that hold in conceptual spaces. We suggest here to enforce an homogeneous proportion T that already holds for completely known features.

Assume we have a Boolean vector incompletely describing a situation with respect to a set of $n + 1$ considered features, say $v = (v_1, ..., v_n, x_{n+1})$, where for simplicity we assume that only x_{n+1} is unknown. For trying to make a plausible guess of the value of x_{n+1}, we have a collection (which may be rather small) of completely informed examples $e^i = (e_1^i, ..., e_n^i, e_{n+1}^i)$ for $i = 1, n$. Then one may have at least three strategies:

i) comparing v to each e^i separately, and using a k-nearest neighbors approach, extending the idea that $T(e, e, e, v)$ should hold true and has $v = e$ as solution.

ii) looking for pairs e^i, e^j such that $T(e_h^i, v_h, v_h, e_h^j)$ makes a continuous homogeneous proportion T for a maximal number of features h, implementing the idea of having v_h between e_h^i and e_h^j ; observe however, that in the Boolean case, this would force to have the trivial situations $T(1,1,1,1)$ or $T(0,0,0,0)$ on a maximal number of features, and tolerate some "approximate" patterns $T(1,1,1,0)$, $T(0,1,1,1)$, $T(0,0,0,1)$, or $T(1,0,0,0)$, while rejecting patterns $T(0,1,1,0)$ and $T(1,0,0,1)$.

iii) looking for triples e^i, e^j, e^k such that $T(e_h^i, e_h^j, e_h^k, v_h)$ makes an homogeneous proportion T for a maximal number of features h.

In cases ii) or iii), the principle amounts to say that if an homogeneous proportion holds for a number of features as great as possible among features h such that $1 \leq h \leq n$, it should still hold for feature $n+1$, which provides an equation for finding x_{n+1} if solvable. If there are several triples that are equally good in terms of numbers of features for which the proportion holds, but lead to different predictions, one may then consider that there is no acceptable plausible value for x_{n+1}.

The application of the first strategy on the above example yields 1 considering that screen 3 is already identical to screen 4 on 3 features. Using the second strategy, we observe that screen 3 is only in "between" screen 2 and screen 4 in the sense described above, leading again to 1 as a solution.

Using the third strategy that should involve 4 distinct items, we can observe that the analogical proportion $A(screen\ 1, screen\ 2, screen\ 4, screen\ 3)$ holds componentwise for features C, R, and D (while it fails with proportions P and I). Again we get 1 as a solution for ensuring an analogical proportion (namely $A(1,1,1,1)$) on S. Observe also that whatever the order in which the screens are considered, an homogeneous proportion holds for features C, R, D, and S. Considering other triples (if available) may lead to other equations having 0 as a solution. A prediction based on the triple making an homogeneous proportion with the incompletely described item on a maximal number of features, should be preferred. In case of ties on this maximal number of features between concurrent triples leading to opposite predictions, no prediction can be given. It is worth noticing that in [1], the use of 0 and 1 in the Boolean coding in their matrices is not just a matter of convention and we cannot exchange the 2 values since it will change the ordering. This is not the case with our approach since A, R, P, I satisfy code independency. The screen example is clearly a toy example but, in [1], similar examples are discussed which could also be handled using homogeneous proportions.

6.3   Analogical proportions in Raven's tests

Among the picture-based IQ tests (the use of pictures avoids the bias of a cultural background), the so-called Raven's Progressive Matrices [24] are considered as a reference for estimating the reasoning component of "the general intelligence". Recently a computational model for solving Raven's Progressive Matrices has been

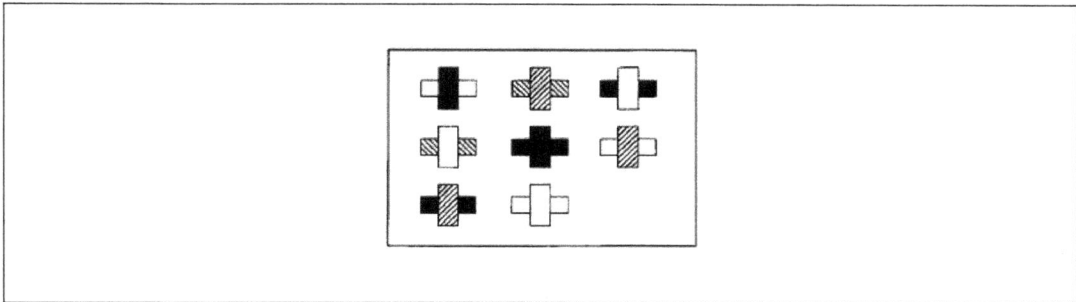

Figure 3: An example of Raven matrix

investigated in [11]. This model combines qualitative spatial representations with analogical comparison via structure-mapping [7]. In the following, we suggest with an example that the Boolean proportion approach can be also used for solving such a test (see [20, 3] for other examples).

Each Raven test is constituted with a 3x3 matrix $pic[i, j]$ of pictures where the last picture $pic[3, 3]$ is missing and has to be chosen among a panel of 8 candidate pictures. An example is given in Figure 3 and its solution in Figure 4. We assume that the Raven matrices can be understood in the following way, with respect to rows and columns:

$$\forall i \in [1, 2], \exists f \text{ such that } pic[i, 3] = f(pic[i, 1], pic[i, 2])$$

$$\forall j \in [1, 2], \exists g \text{ such that } pic[3, j] = g(pic[1, j], pic[2, j])$$

The two complete rows (resp. columns) are supposed to help to discover f (resp. g) and to predict the missing picture $pic([3, 3])$ as $f(pic[3, 1], pic[3, 2])$ (resp. $g(pic[1, 3], pic[2, 3])$).

Obviously, these tests do not fit the standard equation solving scheme, but they follow an extended one telling us that $A((a, b), f(a, b), (c, d), f(c, d))$ holds for lines and $A((a, b), g(a, b), (c, d), g(c, d))$ for columns, i.e.

$$A((pic[1, 1], pic[1, 2]), pic[1, 3], (pic[2, 1], pic[2, 2]), pic[2, 3])$$

$$A((pic[1, 1], pic[2, 1]), pic[3, 1], (pic[1, 2], pic[2, 2]), pic[3, 2])$$

Thus, in that case, we have to consider a pair of cells $(pic[i, 1], pic[i, 2])$ as the first element of an analogical proportion, and then the pair $((pic[i, 1], pic[i, 2]), pic[i, 3])$ provides the 2 first element a and b of the analogical proportion we are considering. In terms of coding, in the example of Figure 3, we may consider the pictures as represented by a pair (or vector) (hr, vr) with one horizontal rectangle hr and a vertical one vr, each of these rectangles having one color among

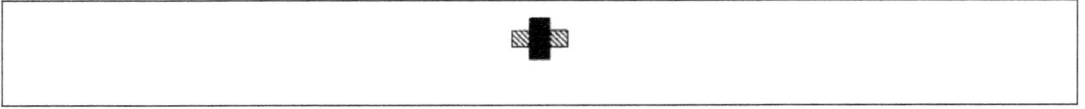

Figure 4: Raven matrix: the solution

Table 19: Raven matrix: a coding

	1	2	3
1	WB	GG	BW
2	GW	BB	WG
3	BG	WW	?i?ii

$Black, White, Grey$, we have then the following obvious encoding of the matrix in Table 19.

It leads to the following analogical patterns (using the traditional notation for analogical proportion $a : b :: c : d$ instead of $A(a, b, c, d)$):

(WB,GG) : BW :: (GW, BB) : WG (1st and 2nd rows)
(WB,GG) : BW :: (BG,WW) : ?i?ii (1st and 3rd rows)
where BW = f(WB,GG) and WG = f(GW,BB).

(WB,GW) : BG :: (GG, BB) : WW (1st and 2nd columns)
(WB,GW) : BG :: (BW,WG) : ?i?ii (1st and 3rd columns)
where BG = f(WB,GW) and WW = g(GG,BB),

or if we prefer, since analogical proportions holds componentwise, we have the following valid proportions
- for the horizontal bars:
(W,G) : B :: (G, B) : W (horizontal analysis)
(W,G) : B :: (B,W) : ?i (horizontal analysis)
(W,G) : B :: (G, B) : W (vertical analysis)
(W,G) : B :: (B,W) : ?i (vertical analysis)
- for the vertical bars:
(B,G) : W :: (W, B) : G (horizontal analysis)
(B,G) : W :: (G,W) : ?ii (horizontal analysis)
(B,W) : G :: (G, B) : W (vertical analysis)
(B,W) : G :: (W,G) : ?ii (vertical analysis)

One can observe that the item (B, W) appears only in the analogical proportions with question marks for horizontal bars, while the items (G, W) and (W, G) appear only in the analogical proportions with question marks for vertical bars. Analogical proportions coming from both horizontal or vertical analysis are insufficient for concluding here. However, we can consider the Raven matrix provides a set of analogical associations without any distinction between those ones coming from the horizontal bars and those ones coming from vertical bars. In other words, we now relax the componentwise reading by considering that what applies to horizontal bars, may apply to vertical bars, and vice-versa. With this viewpoint, it appears that the pair (B, W) and the pair (W, G) are respectively associated to G (vertical association for vertical bar) and B (horizontal association for horizontal bar), which encodes the expected solution GB (as pictured in Figure 4). Note also that (G, W) cannot help predicting ?ii.

6.4 Using heterogeneous proportions "to pick up the one which does not fit"

As it is the case for homogeneous proportions, heterogeneous proportions can also be related to the solving of some type of quiz problem. As we have seen, the truth tables of the heterogeneous proportions highlight a Boolean value (0 or 1) which is different from the 3 remaining ones. It is then natural to think in terms of exception or intruder in a sequence of 4 items: the heterogeneous proportions play a dual role with regard to homogeneous proportions. Given a sequence of objects, they allow to distinguish the object which does not follow the "logic" of the sequence. As a consequence, heterogeneous proportions are suitable for the "Finding the odd one out" problem where a complete sequence of items being given, we have to find the item that does not fit with the other ones and which is, in some sense, an intruder or an anomaly. On this basis, a complete battery of IQ tests has been recently proposed in [8]. Solving 'Find the odd one out' tests (which are visual tests) has been recently tackled in [12] by using analogical pairing between fractal representation of the pictures. It is worth noticing that the approach of these authors takes its root in the idea of analogical proportion. However, this method relies on the use of similarity/dissimilarity measures rather than referring to a formal logical view of analogical proportion. In the following, we show that an opposite type (in some sense) of proportions, namely heterogeneous proportions, provides a convenient way to code and to tackle this problem.

Let us first consider the case of 4 items: obviously, if these items are completely different in many respects, there is no notion of intruder. The intruder comes as soon as there is a kind of unique dissimilarity among an obvious set of similarities or identities. Let us start with a simple case where each item a can be represented as a Boolean vector $a_1, \ldots, a_n$ where n is the number of attributes and $a_i \in \{0, 1\}$. Let us consider the simple example ($lorry, bus, bicycle, car$) (where the obvious intruder is $bicycle$) shown in Figure 5 where $n = 5$ with a straightforward coding.

	canMove	hasEngine	onRoad	has4Wheel	canFly
lorry	1	1	1	1	0
bus	1	1	1	1	0
bicycle	1	0	1	0	0
car	1	1	1	1	0

Figure 5: A simple quiz and its Boolean coding

When considering the item componentwise, we see that:

- for $i = 1, 3, 5$, $H_a(a_i, b_i, c_i, d_i) = H_b(a_i, b_i, c_i, d_i) = H_c(a_i, b_i, c_i, d_i) = H_d(a_i, b_i, c_i, d_i) = 0$.

- for $i = 2, 4$, $H_a(a_i, b_i, c_i, d_i) = H_b(a_i, b_i, c_i, d_i) = H_d(a_i, b_i, c_i, d_i) = 1$.

- for $i = 2, 4$, $H_c(a_i, b_i, c_i, d_i) = 0$

The indexes $1, 3$ and 5 are not useful to pick up the intruder because all the proportions have the same truth value. This is not the case for the indexes 2 and 4: H_a for instance, being equal to 1, insures that there is an intruder (which is not the first element). The intruder is then given by the proportion having the value 0: for instance, $H_c(a_i, b_i, c_i, d_i) = 0$ means that the fact that c is not an intruder is false, which exactly means that c is the intruder for component j. In our example, H_c is 0 on both components 2 and 4: this exactly leads to consider the third element $bicycle$, intruder for the components 2 and 4, as the global intruder. It may be the case that, we do not get the same intruder depending on the component: in that case, a majority vote may be applied and we choose as intruder the one which is intruder for the maximum number of components.

Thanks to the multiple-valued extension, this method can be generalized to the non Boolean case where each item a is represented as a real vector $a_1, \ldots, a_n$ and $a_i \in [0, 1]$. Then, the truth values of H_a, H_b, H_c and H_d on some features may be close to 0, which means that there is no clear intruder according to these features. Let us focus on the other features that are not identical. For each such index j, we can compute the 4 values $H_a(a_j, b_j, c_j, d_j)$, $H_b(a_j, b_j, c_j, d_j)$, $H_c(a_j, b_j, c_j, d_j)$ and $H_d(a_j, b_j, c_j, d_j)$. We know that they cannot be all equal (or close to) 1 since their conjunction is not satisfiable: in fact, exactly one proportion has to be close to

0, thus spotting out the intruder for that component. Applying again a majority vote, we shall consider as global intruder the one which is intruder for the maximum number of components.

In the case where we have to 'Find the odd one out' among more than 4 items, diverse options are available. We may consider all the subsets of 4 items. For each such subset, we apply the previous method to exhibit an intruder (if any). Then the global intruder will be the one which is intruder for the maximum number of subsets.

7 Conclusion

The Boolean modeling of logical proportions which relate 4 items in terms of similarity and dissimilarity, and which may be viewed as a counterpart to numerical proportions, has led to identify a set of 120 distinct proportions. All these logical proportions have the same type of truth table, namely they are true for exactly 6 valuations (and thus false for the 10 remaining valuations). Among this set, only 8 proportions satisfy a so-called code-independency property which makes sure that the evaluation of the proportion remains unchanged when the truth values of the 4 components are changed into their complement (1 is changed into 0, and 0 into 1). This property is important since it ensures that the evaluation of logical proportions will not depend on the positive or negative encoding of the features of the considered items. This set of 8 remarkable proportions divides into 4 homogeneous proportions, and the 4 heterogeneous proportions. These two subsets can be strongly contrasted and appear to be complementary. The 6 valuation patterns that make true homogeneous proportions have all an even number of 1 (and consequently of 0), while for heterogeneous proportions the numbers are odd. Homogeneous proportions are symmetrical, while heterogeneous ones are not. Both types of proportions satisfy remarkable permutation properties. Interestingly enough, these two subsets of logical proportions can be related to two types of IQ tests or quizzes respectively of the type "Find the missing item" and of the type "Find the odd one out". Thus, both from a formal viewpoint and from an applicative viewpoint, heterogeneous proportions appear as a perfect dual of the homogeneous ones. Ultimately, logical proportions provide an elegant unique framework for dealing with IQ tests, from Raven progressive matrices to Find the odd one out quizzes, in a uniform way. Generally speaking, beyond these illustrations, logical proportions still constitute an intriguing set of quaternary connectives, including diverse subsets with remarkable properties, that look interesting for different reasoning purposes where the ideas of similarity and dissimilarity play a role.

References

[1] M. Abraham, D. M. Gabbay, and U. J. Schild. Analysis of the Talmudic argumentum a fortiori inference rule (kal vachomer) using matrix abduction. *Studia Logica*, 92(3):281–364, 2009.

[2] D. Ciucci and D. Dubois. Relationships between connectives in three-valued logics. In S. Greco, B. Bouchon-Meunier, G. Coletti, M. Fedrizzi, B. Matarazzo, and R. R. Yager, editors, *Advances on Computational Intelligence - Proc. 14th Int. Conf. on Information Processing and Management of Uncertainty in Knowledge-Based Systems (IPMU'12), Catania, July 9-13, Part I*, volume 297 of *Communications in Computer and Information Science*, pages 633–642. Springer, 2012.

[3] W. Correa, H. Prade, and G. Richard. When intelligence is just a matter of copying. In L. De Raedt et al., editor, *Proc. 20th Europ. Conf. on Artificial Intelligence, Montpellier, Aug. 27-31*, pages 276–281. IOS Press, 2012.

[4] B. De Finetti. La logique des probabilités. In *Congrès International de Philosophie Scientifique*, pages 1–9, Paris, France, 1936. Hermann et Cie.

[5] D. Dubois and H. Prade. Conditional objects as nonmonotonic consequence relationships. *IEEE Trans. on Systems, Man and Cybernetics*, 24:1724–1740, 1994.

[6] R. M. French. The computational modeling of analogy-making. *Trends in Cognitive Sciences*, 6(5):200 – 205, 2002.

[7] D. Gentner. Structure-mapping: A theoretical framework for analogy. *Cognitive Science*, 7(2):155–170, 1983.

[8] A. Hampshire. The odd one out tests of intelligence. http://www.cambridgebrainsciences.com/browse/reasoning/test/oddoneout, 2010.

[9] S. Klein. Culture, mysticism & social structure and the calculation of behavior. In *Proc. 5th Europ. Conf. in Artificial Intelligence (ECAI'82), Paris*, pages 141–146, 1982.

[10] S. Klein. Analogy and mysticism and the structure of culture (and Com-ments & Reply). *Current Anthropology*, 24 (2):151–180, 1983.

[11] A. Lovett, K. Forbus, and J. Usher. A structure-mapping model of Raven's progressive matrices. In *Proc. of the 32nd Annual Conference of the Cognitive Science Society, Portland, OR*, 2010.

[12] K. McGreggor and A. K. Goel. Finding the odd one out: a fractal analogical approach. In *Proc. 8th ACM Conf. on Creativity and Cognition*, pages 289–298, 2011.

[13] L. Miclet, S. Bayoudh, and A. Delhay. Analogical dissimilarity: definition, algorithms and two experiments in machine learning. *JAIR, 32*, pages 793–824, 2008.

[14] L. Miclet and H. Prade. Handling analogical proportions in classical logic and fuzzy logics settings. In *Proc. 10th Eur. Conf. on Symbolic and Quantitative Approaches to Reasoning with Uncertainty (ECSQARU'09),Verona*, pages 638–650. Springer, LNCS 5590, 2009.

[15] J. Piaget. *Logic and Psychology*. Manchester Univ. Press, 1953.

[16] H. Prade and G. Richard. Analogy, paralogy and reverse analogy: Postulates and

inferences. In *Proc. 32nd Ann. Conf. on Artif. Intellig. (KI 2009), Paderborn*, volume LNAI 5803, pages 306–314. Springer, 2009.

[17] H. Prade and G. Richard. Logical proportions - typology and roadmap. In E. Hüllermeier, R. Kruse, and F. Hoffmann, editors, *Computational Intelligence for Knowledge-Based Systems Design: Proc. 13th Inter. Conf. on Information Processing and Management of Uncertainty (IPMU'10), Dortmund, June 28 - July 2*, volume 6178 of *LNCS*, pages 757–767. Springer, 2010.

[18] H. Prade and G. Richard. Multiple-valued logic interpretations of analogical, reverse analogical, and paralogical proportions. In *Proc. 40th IEEE Inter. Symp. on Multiple-Valued Logic (ISMVL'10), Barcelona, May*, pages 258–263, 2010.

[19] H. Prade and G. Richard. Reasoning with logical proportions. In *Proc. 12th Inter. Conf. on Principles of Knowledge Representation and Reasoning (KR'10), Toronto, Ontario, Canada, May 9-13, (F. Z. Lin, U. Sattler, M. Truszczynski, eds.)*, pages 545–555. AAAI Press, 2010.

[20] H. Prade and G. Richard. Analogy-making for solving IQ tests: A logical view. In *Proc. 19th Inter. Conf. on Case-Based Reasoning, Greenwich, London, 12-15 Sept.*, LNCS, pages 561–566. Springer, 2011.

[21] H. Prade and G. Richard. From analogical proportion to logical proportions. *Logica Universalis*, 7(4), 441–505, 2013.

[22] H. Prade and G. Richard. Homogeneous logical proportions: Their uniqueness and their role in similarity-based prediction. In G. Brewka, T. Eiter, and S. A. McIlraith, editors, *Proc. 13th Inter. Conf. on Principles of Knowledge Representation and Reasoning (KR'12), Roma, June 10-14*, pages 402–412. AAAI Press, 2012.

[23] H. Prade, G. Richard, and B. Yao. Enforcing regularity by means of analogy-related proportions-a new approach to classification. *International Journal of Computer Information Systems and Industrial Management Applications*, 4:648–658, 2012.

[24] J. Raven. The Raven's progressive matrices: Change and stability over culture and time. *Cognitive Psychology*, 41(1):1 – 48, 2000.

[25] S. Schockaert and H. Prade. Interpolation and extrapolation in conceptual spaces: A case study in the music domain. In *Proc. 5th Inter. Conf. on Web Reasoning and Rule Systems (RR'11), Galway, Ireland, Aug. 29-30*. Springer, LNCS, 2011.

[26] A. Tversky. Features of similarity. In *Psychological Review*, number 84, pages 327–352. American Psychological Association, 1977.

Received 1 April 2013

Partially Commutative Linear Logic and Lambek Calculus with Product: Natural Deduction, Normalisation, Subformula Property

Maxime Amblard

Loria (UMR 7503) Université de Lorraine, CNRS, INRIA Nancy Grand-Est
`amblard@loria.fr`

Christian Retoré

LaBRI (UMR 5800) Université de Bordeaux, CNRS
`retore@labri.fr`

Abstract

This article defines and studies a natural deduction system for partially commutative intuitionistic multiplicative linear logic, that is a combination of intuitionistic commutative linear logic with the Lambek calculus, which is non-commutative, and was first introduced as a sequent calculus by de Groote.

In this logic, the hypotheses are endowed with a series-parallel partial order: the parallel composition corresponds to the commutative product, while the series composition corresponds to the noncommutative product. The relation between the two products is that a rule, called *entropy*, allows us to replace a series-parallel order with a sub series-parallel order — this rule (already studied by Retoré) strictly extends the entropy rule initially introduced by de Groote. A particular subsystem emerges when hypotheses are totally ordered: this is Lambek calculus with product, and when orders are empty it is is multiplicative linear logic.

So far only the sequent calculus and cut-elimination have been properly studied. In this article, we define natural deduction with product elimination rules as Abramsky proposed long ago. We then give a brief illustration of its application to computational linguistics and prove normalisation, firstly for the Lambek calculus with product and then for the full partially ordered calculus. We show that normal proofs enjoy the subformula property, thus yielding another proof of decidability of these calculi.

The authors wish thank Jiří Maršík (LORIA, Nancy) for his prompt and efficient rereading.

This logic was shown to be useful for modelling the truly concurrent execution of Petri nets and for minimalist grammars in computational linguistics. Regarding this latter application, natural deduction and the Curry-Howard isomorphism is extremely useful since it leads to the semantic representation of analysed sentences.

Keywords: Logic; Intuitionistic Noncommutative Logic; Lambek calculus; normalisation

1 Presentation

Non-commutative logic arises both as a natural mathematical generalisation of commutative logic and in the modeling of some computational phenomena that require some noncommutativity.

Both truth-value semantics (phase semantics, based on monoids which can be non commutative) and syntax (sequent calculus with sequences rather than sets of formulae, proof nets with non-crossing axiom links) suggest the study of noncommutative linear logic — which fits less well with proof semantics, except for Pomset logic.

Non commutativity is also appealing from a real world application perspective such as in concurrency theory, like the truly concurrent execution of Petri net, and in our favourite application, computational linguistics. This goes back to the fifties and the apparition of the Lambek calculus. We first give a brief presentation of noncommutative logics and then stress their interest with respect to concurrency theory and to computational linguistics, before introducing and studying natural deduction for this calculus.

1.1 Noncommutative linear logics

Linear logic [10] offered a logical view of the Lambek calculus [14] and noncommutative calculi. During many years, the difficulty was to integrate commutative connectives and noncommutative connectives.

The first solution, inspired by denotational semantics (coherence spaces) was Pomset Logic of Retoré (1993) [19, 20]. This logic was defined as an extension of proof net syntax with a faithful interpretation in the category of coherence spaces. In addition to the multiplicative conjunction and disjunction Pomset logic has a noncommutative and self-dual connective called "before". By now it has been generalised and is now studied with extended sequent calculi called Calculus of Structures [12].

Another kind of calculus was introduced as a sequent calculus by de Groote in [8]. Let us stress that it is an intuitionistic calculus (several hypotheses, a single conclusion) and that the classical extension by [3] is quite difficult. The intuitionistic calculus introduced by de Groote, called partially commutative linear logic, consists of a superposition of the Lambek calculus (noncommutative) and of Intuitionistic Linear Logic (commutative). For making a distinction between the two connectives it is necessary that the context includes two different commas that mimic the conjunctions, one being commutative and the other being noncommutative. Hence we deal with series-parallel partial orders over multisets of formulae as right-hand side of sequents. Let us write $(\ldots, \ldots)$ for the parallel composition and $\langle \ldots; \ldots \rangle$ for the noncommutative one: hence $\langle (a, b); (c, d) \rangle$ stands for the finite partial order $a<c, b<c, a<d, b<d$. Of course we would like the two conjunctions to be related. Either the commutative product entails the noncommutative one, or the converse. Surprisingly enough, the two options work just as well — provided one direction is fixed once and for all, of course! This relation between the two products results from a structural rule acting on the order.

$$\frac{\Gamma \text{ ordered by } I \vdash C}{\Gamma \text{ ordered by } J \vdash C}$$

According to our view of this calculus, J can be any order such that $J \subset I$ (ordered are viewed as sets of ordered pairs of formulae in Γ). That's the version of this calculus that we already used for a logical description of Petri net firing, and for viewing derivations in minimalist grammars as proofs. Indeed Bechet, de Groote and Retoré, in [7], showed that only four rewriting rules are needed to obtain all possible series-parallel partial suborders from some series-parallel partial order.

In de Groote's calculus as well as in Abrusci and Ruet's calculus, the resulting order J can only be obtained by replacing some noncommutative commas/products with commutative ones. This is not equivalent to our formulation, indeed Ruet showed in his PhD dissertation [23] that there cannot exist a classical calculus with the classical analogue of our order rule.

Here is an example of a derivation that can be performed in our calculus but not in theirs:

$$\frac{\dfrac{\langle (a, b); (c, d) \rangle \vdash (a \otimes b) \odot (c \otimes d)}{(\langle a; c \rangle, \langle b; d \rangle) \vdash (a \otimes b) \odot (c \otimes d)}}{(a \odot c) \otimes (b \odot d) \vdash (a \otimes b) \odot (c \otimes d)}$$

Up to now our calculus can only be presented with a sequent calculus: there exist neither proof nets, nor natural deduction, only a sequent calculus that has been proven to enjoy cut-elimination in [21]. This is the reason why this paper proposes a natural deduction system, as well as a notion of normal proof, and a normalisation theorem that entails the subformula property.

Commutation of rules is tricky in this calculus. Hence, although we are absolutely convinced that there is a form of confluence and a form of strong normalisation, the possibility to swap these rules results in a lengthy and complicated proof, although there surely is nothing deep into these forgotten proofs.

1.2 Applications of noncommutative (linear) logics

Noncommutativity in logic is rather natural in a resource consumption perspective. A hypothesis is viewed as a resource that can be used but then it is natural to think of how hypotheses are organised and accessible. As argued by Abrusci [2] and others, linearity is a mandatory condition for noncommutativity. The first noncommutative logical calculus, namely the Lambek calculus designed for the grammatical description of natural language, was invented long before linear logic. It is nevertheless a particular system of linear logic, whose relation to other logical calculi, in particular to intuitionistic, was better understood with the help of linear logic.

Concurrency, in which the order of the computations or of the resources matters, is of course a natural application of noncommutative logic(s). In the framework of proofs as programs, and normalisation as computation, Pomset Logic and the subsequent calculus of structures are easier to understand because the order on the conclusions also concerns cuts, which are the computations to be performed [20, 12].

The noncommutative logical calculus that we study in this paper, as well as those by Lambek, Abrusci, Ruet *etc.*, better matches proof search as computation that is at work in linear logic programming [13], planning [16], Petri net firing or parsing in computational linguistics [21]. For instance, Retoré also provided a description of the parallel execution of a Petri net in the calculus we are studying. It is a true concurrency approach, where $a\|b$ is not reduced to $a; b \oplus b; a$ (where $\oplus$ is the nondeterministic choice and ";" sequential composition). An execution according to a series-parallel partial order corresponds to a proof in the partially commutative calculus that we study in this paper; in this order-based approach of parallel computations any set of minimal transitions can be fired simultaneously [21].

Our preferred motivation for such calculi is computational linguistics and grammar formalisms, in particular the deductive description of mildly context-sensitive formalisms. They are assumed to be large enough for natural language constructs, go beyond context-free languages, but admit polynomial parsing algorithms. Logical descriptions of grammar classes as introduced by Lambek are especially appealing because the parse structure and a semantic lexicon automatically lead to a formula which represents the semantics of the analysed sentence. This is especially simple when the Lambek calculus or the partially commutative extensions that we are considering here are described as natural deduction systems. Indeed, the syntactic categories can be turned into semantic categories over two types, individuals (e) and truth values (t), in such a way that the proof in the Lambek calculus (the syntactic analysis) can be turned into a proof in intuitionistic logic, that is a lambda term: when the semantic lambda terms are inserted at the place occupied by the corresponding words, one obtains a lambda term that reduces to the semantics of the sentence, *i.e.* a logical formula written as a lambda term.

Lambek calculus is definitely too restrictive as a syntactic formalism, in particular it only describes context-free languages, thus leaving out many common syntactic constructions. This is the reason to use partially commutative calculi instead of Lambek calculus. In particular Lecomte and Retoré managed to give a logical presentation [15] of Stabler's minimalist grammars [24] in partially commutative linear logic, presented in natural deduction in order to obtain semantic representations of the parsed sentences. In such a framework and for other applications as well, normalisation is quite important : indeed the normal form is the structure of the analysed sentence, and normalisation ensures the coherence of the calculus. The algorithm of normalisation, easily extracted from the proof, is important as well: one define correct sentences as the ones such that some sequent can be proven, and both the parse structure and the semantic reading are obtained from the normal form.

2 Partially Commutative Intuitionnistic Multiplicative Linear Logic (pcIMLL)

2.1 Series-parallel ordered multisets of formulae

Formulae and contexts are defined as in the initial work of de Groote in [8].

Formulae are defined from a set of propositional variables P, by the commutative conjunction ($\otimes$), the noncommutative conjunction ($\odot$), the commutative implication ($\multimap$), the two noncommutative implications (/ and \):

$$\mathrm{L} ::= \mathrm{P} \mid \mathrm{L} \odot \mathrm{L} \mid \mathrm{L} \otimes \mathrm{L} \mid \mathrm{L} \,/\, \mathrm{L} \mid \mathrm{L} \setminus \mathrm{L} \mid \mathrm{L} \multimap \mathrm{L}$$

Contexts, that are left-hand side of sequents, are multisets of formulae endowed with a series-parallel (SP) partial order. Contexts are denoted by upper case Greek letters. Such orders can be defined by two operations: disjoint union, or parallel

$$\dfrac{\Gamma \vdash A \quad \Delta \vdash A \backslash C}{\langle \Gamma ; \Delta \rangle \vdash C} \; [\backslash e] \qquad \dfrac{\Delta \vdash C \,/\, A \quad \Gamma \vdash A}{\langle \Delta ; \Gamma \rangle \vdash C} \; [/e] \qquad \dfrac{\Gamma \vdash A \quad \Delta \vdash A \multimap C}{(\Gamma, \Delta) \vdash C} \; [\multimap e]$$

$$\dfrac{\langle A ; \Gamma \rangle \vdash C}{\Gamma \vdash A \backslash C} \; [\backslash i] \qquad \dfrac{\langle \Gamma ; A \rangle \vdash C}{\Gamma \vdash C \,/\, A} \; [/i] \qquad \dfrac{(A, \Gamma) \vdash C}{\Gamma \vdash A \multimap C} \; [\multimap i]$$

$$\dfrac{\Delta \vdash A \odot B \quad \Gamma[\langle A ; B \rangle] \vdash C}{\Gamma[\Delta] \vdash C} \; [\odot e] \qquad \dfrac{\Delta \vdash A \otimes B \quad \Gamma[(A, B)] \vdash C}{\Gamma[\Delta] \vdash C} \; [\otimes e]$$

$$\dfrac{\Delta \vdash A \quad \Gamma \vdash B}{\langle \Delta ; \Gamma \rangle \vdash A \odot B} \; [\odot i] \qquad \dfrac{\Delta \vdash A \quad \Gamma \vdash B}{(\Delta, \Gamma) \vdash A \otimes B} \; [\otimes i]$$

$$\dfrac{}{A \vdash A} \; [axiom] \qquad \qquad \dfrac{\Gamma \vdash C}{\Gamma' \vdash C} \; [\text{entropy — whenever } \Gamma' \sqsubset \Gamma]$$

Figure 1: Rules of pcIMLL.

composition, denoted by (Γ, Δ) and series composition denoted by $\langle \Gamma ; \Delta \rangle$: the domain is the disjoint union of Γ and Δ, and every formula in Γ comes before any formula in Δ. Contexts obey the following syntax:

$$\text{CTX} ::= \text{L} \,|\, \langle \text{CTX} ; \text{CTX} \rangle \,|\, (\text{CTX}, \text{CTX})$$

For example, the context $\langle \langle B ; (A \multimap (B \backslash (D \,/\, C), A)) \rangle ; C \rangle$ denotes the sp order, $Succ(B) = (A, A \multimap (B \backslash (D \,/\, C)))$, $Succ(A) = Succ(A \multimap (B \backslash (D \,/\, C))) = \{C\}$ where $Succ(X)$ is the multiset of the immediate successors of X (we only consider finite orders).

The term denoting an sp order is unique up to the commutativity of $(\_, \_)$ and to the associativity of both $(\_, \_)$ and $\langle \_ ; \_ \rangle$. The term notation is only a short hand: even though the sp terms denoting them are different, the left-hand sides of two sequents are considered equal whenever they are equal as partially ordered multisets.

An expression $\Gamma[*]$ stands for a context in which we distinguish a specific element $[*]$ and an expression $\Gamma[\Delta]$ denotes the context obtained be replacing $*$ in $\Gamma[*]$ by the context Δ in $\Gamma[*]$.

Figure 1 shows the rules of pcIMLL. It uses the standard rules of commutative multiplicative intuitionistic linear logic and of non-commutative multiplicative intuitionistic linear logic. Both have introductions and eliminations for the implicative connectives and the product connectives. In addition there is the axiom rule and the entropy rule ($\sqsubset$) which correspond to inclusions of orders.

This calculus deserves some explanation and comments:

Entropy $\Gamma' \sqsubset \Gamma$ whenever these contexts that are sp partially ordered multisets of formulae have the same multiset domain $|\Gamma| = |\Gamma'|$, and whenever considering each occurrence of a formula as distinct if $A < B$ in Γ' then $A < B$ in Γ as well. The inclusion $\sqsubset$ of series-parallel partial orders can be viewed as a rewriting system (modulo commutativity and associativity) on the sp term denoting them as shown in [7] – see also [21] where the order rule is used the other way round, but as said in the introduction, it does not change normalisation.

Product elimination rules In the $\otimes$ and $\odot_e$ rules, A and B must be equivalent:

$$\forall X \neq A, B \begin{cases} X < A \Leftrightarrow X < B \\ X > A \Leftrightarrow X > B \end{cases}$$

In the $\otimes_e$ case A and B are equivalent and uncomparable while in the $\odot_e$ case, they are equivalent and $A < B$. In the conclusions of rules, they are replaced by the context which lead to $A \otimes B$ (*i.e.* Δ in figure 1). Abramsky introduced similar rules long ago in [1].

Although we do have normalisation and the subformula property (see next sections) we avoided complicated rules of the kind introduced in [18] for MLL. We assume her rules are motivated by other properties; indeed, they work for the complete linear calculus with additive and exponentials. But in the restricted multiplicative case, our simple rules are preferable.

2.2 Principal branches

Let δ be a proof of this calculus, and let S_j be an occurrence of a sequent $|S_j|$ in δ, and $|S_j|^r$ be the conclusion of this sequent, that is the unique formula in the right-hand side of this sequent.

We write $B(S_0)$ for the **principal branch** starting with the sequent S_0 — see e.g. [11, p. 75] or [17, p. 35]. It is the smallest path which contains S_0 and is closed under the following operations:

1. If $S \in B(S_0)$ is obtained by a unary rule R from an occurrence of a sequent S', then $S' \in B(S_0)$.

2. If $S \in B(S_0)$ is obtained by a product elimination $\odot_e$ (*resp.* $\otimes_e$), then the premise $|S'| = \Gamma[\langle A, B\rangle] \vdash C$ (*resp.* $|S'| = \Gamma[(A, B)] \vdash C$) is also in $B(S_0)$.

3. If $S \in B(S_0)$ is obtained by an implicative elimination rule $\backslash_e$ (*resp.* $/_e$, $\multimap_e$), then the premise $|S'| = \Delta \vdash A \backslash C$ (*resp.* $|S'| = \Delta \vdash A / C$, $|S'| = \Delta \vdash A \multimap C$) is also in $B(S_0)$.

For every path in a principal branch $B(S)$ from S to S_i such that $|S|^r = |S_i|^r$, if $|S|$ is the conclusion of an elimination rule and $|S_i|$ the conclusion of introduction rule, the two sequents are said to be **conjoined** — if there are no rules in-between these two these two rules, they define a redex.

3 A brief example using pcIMLL in Computational Linguistics: categorial minimalist grammars

Before we prove the normalisation and subformula property of pcIMLL, let us illustrate briefly our use of this calculus in computational linguistics — for more details see [15, 4, 6]. As said above, Lambek calculus is too restricted to describe some constructions in natural language syntax, hence, we need a richer logical calculus, in order to remain in the parsing-as-deduction paradigm and to have a simple syntax semantic interface. The calculus presented in this paper, namely partially commutative intuitionistic multiplicative linear logic, pcIMLL, is able to account of many more syntactical phenomena, especially when viewed as a natural deduction system.

Indeed, this logical calculus can account for Stabler's minimalist grammars, which are an elegant and computationally efficient formalisation of Chomsky's minimalist program. As first observed in [9], such a view of syntax is not too far from categorial grammar like Lambek grammars. [24, 22, 4, 6]. Minimalist grammars cover all (or most of) syntactic constructions, but lacks a simple connection to semantics because it is not a deductive system.

Therefore we defined *categorial minimalist grammars*, a deductive grammatical formalism that resembles Lambek grammars: a lexicon maps every word into a formula of partially commutative linear logic which describes its interaction with other words. The main difficulty was to mimic the Chomskyan notion of movement: the proper word order is recovered in a second step that labels the proof nodes. We do not use all the possible combinations of rules of pcIMLL but only combination of fixed sequences of rules, that correspond to minimalist-grammar rules. Thus parse structures are derivations in pcIMLL where every formula of every sequent in the proof is labelled with strings of words and variables. Parsing consists in deriving in natural deduction *sentence* : C from axioms $x : A \vdash x : A$ and proper axioms $\vdash w : T$ where T is the formula associated with w in the lexicon.

The MERGE rule is almost like residuation rules in categorial grammars with *noncommutative* implications:

$$\frac{\vec{x}:\Delta \vdash w:A \quad \vec{x}':\Gamma \vdash w':A \setminus C}{\dfrac{\vec{x}:\Delta; \vec{x}':\Gamma \vdash ww':C}{\vec{x}:\Delta, \vec{x}'\Gamma \vdash ww':C}\,[entropy]}\,[\setminus e] \qquad \text{or} \qquad \frac{\vec{x}':\Gamma \vdash w':C/A \quad \vec{x}:\Delta \vdash w:A}{\dfrac{\vec{x}':\Gamma; \vec{x}:\Delta \vdash w'w:C}{\vec{x}':\Gamma; \vec{x}:\Delta \vdash w'w:C}\,[entropy]}\,[/e]$$

The MOVE rule was trickier to encode and required the *commutative* product:

$$\frac{\vec{y}:\Gamma \vdash s:A \otimes B \quad \vec{x}:\Delta, x:A, y:B, \vec{x}':\Delta' \vdash t:C}{\vec{x}:\Delta, \vec{y}:\Gamma, \vec{x}':\Delta' \vdash t[s\,/\,x, \epsilon\,/\,y]:C}\,[\otimes_e]$$

The MOVE rule is typical of Chomskyan linguistics: our encoding in pcIMLL mimics the movement of the constituent / string s from the place y to the place x .

Here is a simple example involving movement with a few entries from an Italian lexicon — null subjects are allowed in this language, it makes the example simpler. Observe how an interrogative noun phrase is moved to the leftmost position:

| que | $wh \otimes (k \otimes d)\,/\,n$ | cosa | n | ϵ | $k \otimes d$ |
| fai | $k \setminus d \setminus v\,/\,d$ | infl | $k \setminus t\,/\,v$ | comp | $wh \setminus c/t$ |

Figure 2: Analysis of "che cosa ϵ fai ?"

Our interest for such analyses is to automatically compute the semantic representation (usually, a formula of first-order or higher-order logic) with correct quantifier scope. The semantic representation of our example is:

$$\exists?(\lambda x\,(\wedge(cosa(x))(far(tu,x)))) \equiv \exists?x(cosa(x) \wedge far(tu,x))$$

Our presentation of categorial minimalist grammars in pcIMLL is extremely brief and sketchy, but it gives an idea of why the logical calculus in this paper and its natural deduction formulation is relevant to computational linguistics. Details can be found in [15, 4, 5].

Regarding the application to true concurrency and Petri nets, see [21]

4 Normalisation of Lambek calculus with product (L$_\odot$)

Before to give an algorithm for normalisation for the full pcIMLL calculus, let us deal with L$_\odot$, that is the Lambek calculus with product. For this calculus we shall define normal natural deductions, give a normalisation algorithm and prove that it always reach a normal proof. This restricted case is simpler, in particular *entropy* rule is never used. Nevertheless this simpler case shows the difficulties of rule commutations in such calculi, which will be tougher for full pcIMLL in the next section.

4.1 Some properties of the Lambek calculus with product L$_\odot$

Lambek calculus with product (L$_\odot$) is the restriction of pcIMLL to the connectives: \, / and $\odot$. Furthermore, contexts are always totally ordered multi sets of formulae, that are sequences of formulae: hence the associative $\langle ...; ...\rangle$ braces are omitted. Observe that in this setting, the entropy rule cannot be applied.

Property 1. *Let R be a product elimination rule $\odot_e$ yielding $\Gamma[\Delta] \vdash C$ from a proof δ_0 of $\Delta \vdash A \odot B$ and a proof of $\Gamma[A, B] \vdash C$ obtained by a rule R' from a proof δ_1 of $\Theta[A, B] \vdash X$ and from a second proof δ_2 of $\Psi \vdash U$ — if R' is a binary rule.*

A proof for the same sequent $\Gamma[\Delta] \vdash C$ can be derived by the following two steps:

1. *apply R between the proof δ_0 of $\Delta \vdash A \odot B$ and the proof δ_1 of $\Theta[A, B] \vdash X$ as conclusion, yielding $\Theta[\Delta] \vdash X$;*

2. *apply R' to this new proof and to the proof δ_2.*

In other words, provided that the hypotheses A, B that are cancelled by the R product elimination rule are in the same premise of the R' rule, the R product elimination rule can swing over R', as shown in figure 3.

Proof. The proof of this proposition is rather lengthy but simple: it is a case study of all the possible rules R' above the product elimination R. Here are the possible cases:

  ∘ R' is $\backslash_e$:

$$
\begin{array}{c}
\vdots \delta_0 \qquad \dfrac{\Psi \vdash U \quad \Theta[A,B] \vdash X}{\Gamma[A,B] \vdash C} \ R' \\[2pt]
\dfrac{\Delta \vdash A \odot B \qquad \qquad \qquad}{\Gamma[\Delta] \vdash C} \ R
\end{array}
\quad \Rightarrow \quad
\begin{array}{c}
\vdots \delta_2 \qquad \dfrac{\Delta \vdash A \odot B \quad \Theta[A,B] \vdash X}{\Theta[\Delta] \vdash X} \ R \\[2pt]
\dfrac{\Psi \vdash U \qquad \qquad \qquad}{\Gamma[\Delta] \vdash C} \ R'
\end{array}
$$

Figure 3: A product elimination rule R swings over a rule R' in $L_\odot$.

- hypotheses A, B are in the left premise of $\backslash_e$:

$$
\dfrac{\Gamma \vdash A \odot B \quad \dfrac{A, B \vdash D \quad \Delta \vdash D \backslash C}{A, B, \Delta \vdash C} \ [\backslash_e]}{\Gamma, \Delta \vdash C} \ [\odot_e]
$$

$$
\Rightarrow \quad \dfrac{\dfrac{\Gamma \vdash A \odot B \quad A, B \vdash D}{\Gamma \vdash D} \ [\odot_e] \quad \Delta \vdash D \backslash C}{\Gamma, \Delta \vdash C} \ [\backslash_e]
$$

- hypotheses A, B are in the right premise of $\backslash_e$:

$$
\dfrac{\Gamma \vdash A \odot B \quad \dfrac{\Delta \vdash D \quad A, B \vdash D \backslash C}{\Delta, A, B \vdash C} \ [\backslash_e]}{\Delta, \Gamma \vdash C} \ [\odot_e]
$$

$$
\Rightarrow \quad \dfrac{\Delta \vdash D \quad \dfrac{\Gamma \vdash A \odot B \quad A, B \vdash D \backslash C}{\Gamma \vdash D \backslash C} \ [\odot_e]}{\Delta, \Gamma \vdash C} \ [\backslash_e]
$$

○ R' is $/_e$ — symmetrical to the previous case.

○ R' is $\backslash_i$:

$$
\dfrac{\Gamma \vdash A \odot B \quad \dfrac{D, \Delta, A, B, \Delta' \vdash C}{\Delta, A, B, \Delta' \vdash D \backslash C} \ [\backslash_i]}{\Delta, \Gamma, \Delta' \vdash D \backslash C} \ [\odot_e]
$$

63

$$\Rightarrow \quad \frac{\Gamma \vdash A \odot B \quad D, \Delta, A, B, \Delta' \vdash C}{\dfrac{D, \Delta, \Gamma, \Delta' \vdash C}{\Delta, \Gamma, \Delta' \vdash D \backslash C} \,[\backslash_i]}\,[\odot_e]$$

○ R' is $/_i$ — symmetrical to the previous case.

○ R' is $\odot_e$:

• hypotheses A, B are in the left premise of R':

$$\frac{\Gamma \vdash A \odot B \quad \dfrac{\Delta, A, B, \Delta' \vdash C \odot D \quad \Phi, C, D, \Phi' \vdash E}{\Phi, \Delta, A, B, \Delta', \Phi' \vdash E}\,[\odot_e]}{\Phi, \Delta, \Gamma, \Delta', \Phi' \vdash E}\,[\odot_e]$$

$$\Rightarrow \quad \frac{\dfrac{\Gamma \vdash A \odot B \quad \Delta, A, B, \Delta' \vdash C \odot D}{\Delta, \Gamma, \Delta' \vdash C \odot D}\,[\odot_e] \quad \Phi, C, D, \Phi' \vdash E}{\Phi, \Delta, \Gamma, \Delta, \Phi' \vdash E}\,[\odot_e]$$

• hypotheses A, B are in the right premise of R':

$$\frac{\Gamma \vdash A \odot B \quad \dfrac{\Delta \vdash C \odot D \quad \Phi, A, B, C, D, \Phi' \vdash E}{\Phi, A, B, \Delta, \Phi' \vdash E}\,[\odot_e]}{\Phi, \Gamma, \Delta, \Phi' \vdash E}\,[\odot_e]$$

$$\Rightarrow \quad \frac{\Delta \vdash C \odot D \quad \dfrac{\Gamma \vdash A \odot B \quad \Phi, A, B, C, D, \Phi' \vdash E}{\Phi, \Gamma, C, D, \Phi' \vdash E}\,[\odot_e]}{\Phi, \Gamma, \Delta, \Phi' \vdash E}\,[\odot_e]$$

○ R' is $\odot_i$:

• hypotheses A, B are in the left premise of R':

$$\frac{\Gamma \vdash A \odot B \quad \dfrac{\Delta, A, B, \Delta' \vdash C \quad \Phi \vdash D}{\Delta, A, B, \Delta', \Phi \vdash C \odot D}\,[\odot_i]}{\Delta, \Gamma, \Delta', \Phi \vdash C \odot D}\,[\odot_e]$$

$$\Rightarrow \quad \dfrac{\dfrac{\Gamma \vdash A \odot B \quad \Delta, A, B, \Delta' \vdash C}{\Delta, \Gamma, \Delta' \vdash C} \, [\odot_e] \quad \Phi \vdash D}{\Delta, \Gamma, \Delta', \Phi \vdash C \odot D} \, [\odot_i]$$

- hypotheses in the right premise of R':

$$\dfrac{\Gamma \vdash A \odot B \quad \dfrac{\Delta \vdash C \quad \Phi, A, B, \Phi' \vdash D}{\Delta, \Phi, A, B, \Phi' \vdash C \odot D} \, [\odot_i]}{\Delta, \Phi, \Gamma, \Phi' \vdash C \odot D} \, [\odot_e]$$

$$\Rightarrow$$

$$\dfrac{\Delta \vdash C \quad \dfrac{\Gamma \vdash A \odot B \quad \Phi, A, B, \Phi' \vdash D}{\Phi, \Gamma, \Phi' \vdash D} \, [\odot_e]}{\Delta, \Phi, \Gamma, \Phi' \vdash C \odot D} \, [\odot_i]$$

All possible cases of combinations of rules have been examined. The product elimination has the ability to swing over any rule provided the cancelled hypotheses are in the same premise. □

Definition 1. *Let R be a product elimination rule:*

$$\dfrac{\begin{matrix} \vdots \, \delta_0 \\ \Delta \vdash A \odot B \end{matrix} \quad \dfrac{\begin{matrix} \vdots \, \delta_2 \\ \Gamma_2 \vdash C_2 \end{matrix} \quad \begin{matrix} \vdots \, \delta_1 \\ \Gamma_1 \vdash C_1 \end{matrix}}{\Gamma[A, B] \vdash C} \, R'}{\Gamma[\Delta] \vdash C} \, R$$

*A $\odot_e$ rule R cancelling the two hypotheses A and B is said to be **as high as possible** if R' is binary, and A and B are not in the same sequent, i.e. A is in Γ_2 and B is in Γ_1.*

As usual, a **redex** consists in an introduction rule of a given connective immediately followed by the elimination rule of the same connective. Hence, in this calculus, there are four *redexes*: one for $/$, one for $\backslash$ and two for $\odot$ (depending in which premise it takes place). The reductions patterns are given below. For simplicity we only write the conclusions of the sequents and leave out the contexts: this is unambiguous given that contexts are plain sequences of formulae.

- Redex$_/$: $/_i$ immediately followed by $/_e$.

65

$$
\begin{array}{c}
[D]_1 \\
\vdots\, \delta_0 \\
C \\
\hline
C\,/\,D \;\; [/_i]_1 \qquad \vdots\, \delta_1 \\
\hline
\qquad\qquad\qquad D \\
\hline
C \;\; [/_e]
\end{array}
\qquad \Rightarrow \qquad
\begin{array}{c}
\vdots\, \delta_1 \\
D \\
\vdots\, \delta_0 \\
C
\end{array}
$$

○ Redex$_\backslash$: $/_i$ immediately followed by $/_e$: symmetrical.

$$
\begin{array}{c}
\qquad\qquad [D]_1 \\
\qquad\qquad \vdots\, \delta_0 \\
\vdots\, \delta_1 \qquad C \\
\hline
D \qquad D\,\backslash\,C \;\; [\backslash_i]_1 \\
\hline
C \;\; [\backslash_e]
\end{array}
\qquad \Rightarrow \qquad
\begin{array}{c}
\vdots\, \delta_1 \\
D \\
\vdots\, \delta_0 \\
C
\end{array}
$$

○ Redex$_\odot$: introduction $\odot_i$ (left) immediately followed by elimination $\odot_e$.

$$
\begin{array}{c}
\vdots\, \delta_1 \quad \vdots\, \delta_2 \\
A \qquad B \\
\hline
A \odot B \;\; [\odot_i] \qquad\qquad\qquad\qquad D \\
\hline
D \;\; [\odot_e]_1
\end{array}
\qquad
\begin{array}{c}
[A]_1 \quad [B]_1
\end{array}
\qquad \Rightarrow \qquad
\begin{array}{c}
\vdots\, \delta_1 \quad \vdots\, \delta_2 \\
A \qquad B \\
\vdots \\
D
\end{array}
$$

○ Redex$_\odot$: introduction $\odot_i$ (right) immediately followed by elimination $\odot_e$.

$$
\begin{array}{c}
\vdots\, \delta_1 \qquad A \quad B \\
\qquad\qquad \hline \\
A \odot B \qquad A \odot B \;\; [\odot_i] \\
\hline
A \odot B \;\; [\odot_e]
\end{array}
\qquad \Rightarrow \qquad
\begin{array}{c}
\vdots\, \delta_1 \\
A \odot B
\end{array}
$$

From the notion of redex, we define a generalisation that we call a **k-extended-redex**.

Definition 2. *Every path of a principal branch $B(S_0)$ of length k from S_0 to S_n with $|S_0|^r = |S_n|^r$ (the conclusions of those two sequents are the same), such that $|S_0|$ is the conclusion of an elimination rule R_e and S_n is the conclusion of an introduction rule R_i of the same connective, is called a **k-extended-redex**. Note that 0-extended-redexes are redexes.*

Proposition 1. *A k-extended-redex only contains $\odot_e$ rules or a proper sub k'-extended redex, with $k' < k$.*

Proof. Assume that one of the occurrences of X results from a $/_e$ rule between X / U, and U. In this case, the k-extended-redex contains a smaller k'-extended-redex, between this elimination and same the introduction rule. Otherwise only $\odot_e$ rules preserve the conclusion of the sequent, and they can be used an unspecified number of times without changing the conclusion of the sequent (*i.e.* X). $\qquad\square$

4.2   Normalisation of $\mathbf{L}_\odot$

Definition 3. *A **normal proof** is a proof which contains no k-extended-redexes and where all $\odot_e$ rules are as high as possible.*

Given a proof δ, and $PER(\delta)$ its $\odot_e$ rules, we define for a product elimination rule R in $PE(\delta)$ the two following integers:

1. $g(R)$ which is k if there is a k-extended-redex in the principal branch $B(S_0)$ starting from the conclusion of R and 0 otherwise.

2. the integer $d_{conj}(R)$ is the number of rules other than product elimination rules (*i.e.* not in $PER(\delta)$) between R and the rule which gathers in the same sequent the hypotheses A and B that are cancelled by R.

With those measures, we define the measure of a proof that will be used to established the normalisation. It is a lexically ordered triple of integers:

$$|\delta| = \langle n(\delta), h(\delta), g(\delta) \rangle$$

where:

- $n(\delta)$ is the number of rules of δ. The number of rules decreases when one reduces a redex as in any linear calculus.

- $h(\delta) = \sum_{R \in PER(\delta)} d_{conj}(R)$ when PER is the set of Proudct Elimination rules. The number $h(\delta)$ is 0 when all product elimination rules are as high as possible (cf. definition 1).

- $g(\delta) = \min_{R \in PER(\delta)}(g(R))$ which is 0 if and only if δ contains no k-extended-redex (cf. definition 3).

Property 2. *A proof δ is normal if and only if it does not contain 0-extended-redex, and $h(\delta) = g(\delta) = 0$.*

Proof. Let δ a proof of $L_\odot$,

- $h(\delta)$ is there to make sure that $\odot_e$ rule reach their highest possible position. During this phase, k-extended-redexes$_/$ and $\backslash$ may appear and are reduced afterwards. If every $\odot_e$ is as high as possible then $h(\delta) = 0$. Only $\odot$ k-extended-redexes may still exist in δ. This case is presented in example 1 of Figure 4.

- $g(\delta)$ represents the number of rules in a $\odot$ k-extended-redex. When it is zero, there is no more k-extended-redex$\odot$ in δ. This case is presented in example 2 of Figure 4.

□

$$g(\backslash_e) = 1 \text{ (1-extended-redex)}$$

$$\cfrac{\vdash C \quad \cfrac{\vdash E \odot F \quad \cfrac{\cfrac{C, E, F \vdash A / B}{E, F \vdash C \backslash (A / B)} [\backslash_i]}{\vdash C \backslash (A / B)} [\odot_e]}{\vdash A / B} [\backslash_e]}{}$$

$$h(\odot_e) = 1 \text{ (}\odot_e \text{ is not as high as possible)}$$

$$\cfrac{\vdash E \odot F \quad \cfrac{\vdash C \quad \cfrac{\cfrac{E \vdash (C \backslash A) / B \quad F \vdash B}{E, F \vdash C \backslash A} [/_e]}{E, F \vdash A} [/_e]}{\vdash A / B} [\odot_e]}{}$$

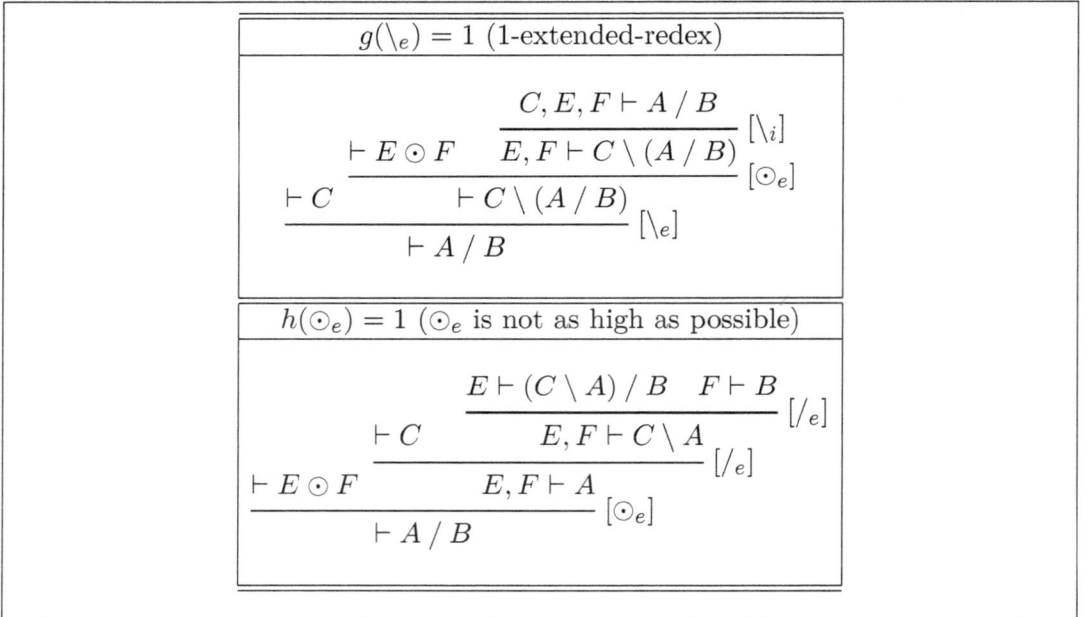

Figure 4: Examples of proof which are not in normal form

Theorem 1. *Every proof δ in $L_\odot$ calculus can be turned into a normal form which is unique.*

Proof. We proceed by induction on $|\delta|$. By induction hypotheses, every proof δ' of size $|\delta'| < \langle r, d, g \rangle$ has a unique normal form. Given a proof δ of size $|\delta| <= \langle r, d, g \rangle$, let us show that δ has a unique normal form as well.

1. If δ contains a redex, we can reduce it, and let δ', be the reduced proof. We have $n(\delta') < n(\delta)$, hence $|\delta'| < \langle r, d, g \rangle$ and by induction, δ' has a unique normal form, and therefore so does δ.

2. Else, if δ contains no redex,

   (a) If $d \neq 0$: let R be the lowest $\odot_e$ rule such that $d(R) \neq 0$. Hence, there exists a rule $R' \neq \odot_e$ higher than R, and R can be lifted above all $\odot_e$ and finally R can swing over R'. The induced proof δ' is such that $n(\delta') = n(\delta)$ and $h(\delta') = h(\delta) - 1$. The number $d_{conj}(R_i)$, for an $\odot_e$ rule R_i below R, is zero because R does not contribute to $d_{conj}(\_)$. Therefore $\delta' < \langle r, d, g \rangle$, and by induction , δ' has a unique normal form, hence so does δ.

   (b) Else:

      i. If $g \neq 0$: let R' such that $g(R') = g$. This rule can swing over its left premise. The number of rules and the sum remain unchanged. In its part, the places of the $\odot_e$ rules are just exchanged, hence g decrease of 1. Hence the proof δ' is such that $|\delta'| < |\delta|$. By induction, δ' has a unique normal form, hence so does δ.

      ii. Else: using the property 2, the proof is in normal form.

Consequently an $\odot_e$ rule R cancelling two hypotheses A and B may only appear below the binary rule R_b which gathers the two hypothesis A and B in one sequent. Moreover, only one $\odot_e$ can be "as high as possible", *i.e.* immediately below the binary rule R_b. Indeed, an $\odot_e$ rule can only cancel two adjacent free hypotheses: the rightmost hypothesis of the left premise of R_b and the leftmost hypothesis of the right premise os R_b. We thus assign a *unique* position to each $\odot_e$ rule, and therefore the normal form is unique.

$\square$

All proofs have a unique normal form which can be computed using the aforementioned strategy. The normal forms of the two previous examples, figure 4, are the two following proofs:

4.3   Subformula property for $\mathbf{L_\odot}$

Theorem 2. *Any normal proof δ of the Lambek calculus with product of a sequent $\Gamma \vdash C$ satisfies the subformula property: every formula in δ is a subformula of some hypothesis in Γ or of the conclusion C.*

Proof. Here we prove a stronger result than the plain subformula property:

$g(\backslash_e) = 0$ (0-extended-redex, *i.e.* visible redex to be reduced)

$$\dfrac{\vdash E \odot F \qquad \dfrac{\vdash C \qquad \dfrac{C, E, F \vdash A \,/\, B}{E, F \vdash C \backslash (A \,/\, B)}[\backslash_i]}{\vdash A \,/\, B}[\backslash_e]}{\vdash C \backslash (A \,/\, B)}[\odot_e]$$

$h(\odot_e) = 0$ ($\odot_e$ is by now as high as possible)

$$\dfrac{\vdash C \qquad \dfrac{\vdash E \odot F \qquad \dfrac{E \vdash (C \backslash A) \,/\, B \quad F \vdash B}{E, F \vdash C \backslash A}[/_e]}{\vdash C \backslash A}[\odot_e]}{E, F \vdash A}[/_e]$$

Figure 5: Applying a generalised reduction step to the examples of figure 4

1. every formula in a normal sub-proof is a formula of some hypotheses or of the conclusion of the proof ;
2. and if the last rule used is an $\backslash_e$ or $/_e$ every subformula is a subformulae of some (uncancelled) hypothesis.

We proceed by a standard induction on the height of the proof, according to the nature of the last rule:

1. A proof consisting in an axiom clearly enjoys the two properties.
2. If the last rule R is $\backslash_e$: let δ be the following proof, where Γ_i is the set of hypotheses used in the sub-proof δ_i, for $i \in [1, 2]$:

$$\dfrac{\begin{matrix}\Gamma_1 \\ \vdots\ \delta_1 \\ C\end{matrix} \qquad \dfrac{\begin{matrix}\Gamma_2 \\ \vdots\ \delta_2\end{matrix}}{C \backslash D}[R]}{D}[\backslash_e]$$

Using the induction hypothesis:

- In δ_1 every formula is subformula of C or Γ_1;

- In δ_2 every formula is subformula of $C \setminus D$ or Γ_2.

The conclusion D and the premise C are direct subformulae of the premise $C \setminus D$. We have to consider the rule $[R]$ above this premise:

- if R is $/_e$ or $\setminus_e$: we use the induction hypothesis, we conclude that $C \setminus D$ is a subformula of Γ_2. Then every formula of δ is a subformula of Γ_2.

- if R is $\setminus_i$: it is impossible because the rule should be a 0-extended-redex, while δ is in normal form.

- if R is $/_i$: this case is structurally impossible because this rule cannot derive $C \setminus D$.

- if R is $\odot_i$: this case is also impossible because this rule cannot derive $C \setminus D$.

- if R is $\odot_e$. Once more, it depends on the rule R' above R:

$$
\cfrac{
\begin{array}{c}
\Gamma_1 \\
\vdots \; \delta_1 \\
\vdots \\
C
\end{array}
\qquad
\cfrac{
A \odot B
\qquad
\cfrac{
\begin{array}{c}
\Gamma_2[A, B] \\
\vdots \\
\vdots \; \delta_2 \\
\vdots
\end{array}
}{C \setminus D} \; [R']
}{C \setminus D} \; [\odot_e]
}{D} \; [\setminus_e]
$$

- If R' is $\setminus_e$ or $/_e$, by the induction hypothesis, $C \setminus D$ is a subformula of some hypotheses.

- If R' is $\setminus_i$: impossible because it would result in a 1-extended-redex, while δ is in normal form.

- If R' is one of the other introduction rules ($\setminus_i$ or $\odot_i$): these cases are structurally impossible since these rules cannot derive $C \setminus D$.

- If R' is $\odot_e$, once more, we must consult the rule above. As the number of rules above a given rule is finite, the proof contains a sequence of rules that necessarily matches the following pattern:

71

$$
\cfrac{
  \cfrac{
    \Gamma_1 \\ \vdots\ \delta_1 \\ C
  }{}
  \quad
  \cfrac{
    A_1 \odot B_1 \quad
    \cfrac{
      \cfrac{
        A_n \odot B_n \quad
        \cfrac{
          \cfrac{\Gamma_2[A_1, \cdots, A_n, B_n, \cdots, B_1] \\ \vdots\ \delta_2}{C \setminus D}\ [R'']
        }{}
      }{C \setminus D}\ [\odot_e] \\
      \vdots \\
      C \setminus D
    }{C \setminus D}\ [\odot_e]
  }{C \setminus D}\ [\backslash_e]
}{D}
$$

Then one of the following case applies:

* There are only $\odot_e$ rules in this sequence of rules, and $C \setminus D$ is one of the hypotheses.

* Otherwise the path stops on a rule R^n which according to what we said about R' above, can only be $/_e$ or $\backslash_e$: therefore, by induction hypothesis $C \setminus D$ is subformula of one of the hypotheses.

In every case, the conclusion of $\backslash_e$ is a subformula of the hypotheses and the property holds.

3. R is $/_e$: this case is similar to R is $\backslash_e$.

4. R is $\backslash_i$: let δ be the following proof, where Γ_1 is the set of hypotheses used in the sub-proof δ_1:

$$
\cfrac{
  \cfrac{
    C, \Gamma_1 \\ \vdots\ \delta_1 \\ D
  }{}
}{C \setminus D}\ [\backslash_i]
$$

In δ_1 every formula is a subformula of D or of C and Γ_1. Furthermore, D is a subformula of $C \setminus D$. Then, every formula of δ is subformula of C, Γ_1 or $C \setminus D$.

5. R is $/_i$: is symmetrical to the previous case.

6. R is $\odot_i$: let δ be the following proof, where Γ_i is the set of hypotheses used in the sub-proof δ_i, for $i \in [1, 2]$:

$$\frac{\begin{array}{cc} \Gamma_1 & \Gamma_2 \\ \vdots\, \delta_1 & \vdots\, \delta_2 \\ C & D \end{array}}{C \odot D}\,[\odot_i]$$

- In δ_1 every formula is a subformula of C or of Γ_1.
- In δ_2 every formula is a subformula of D or of Γ_2.

Furthermore, C and D are subformulae of $C \odot D$. Hence, every formula of δ is subformula of Γ_1, Γ_2 or of $C \odot D$.

7. R is $\odot_e$: let δ be the following proof, where Γ_i is the set of hypotheses used in the sub-proof δ_i, for $i \in [1,2]$:

$$\frac{\begin{array}{cc} \Gamma_1 & \Gamma_2 \\ \vdots\, \delta_1 & \vdots\, \delta_2 \\ A \odot B & D \end{array}}{D}\,[\odot_e]$$

- In δ_1, every formula is a subformula of $A \odot B$ or of Γ_1.
- In δ_2, every formula is a subformula of D or of Γ_2.

The conclusion of δ is the conclusion of one premise, hence the property holds for the part of the proof which the conclusion belongs to, *i.e.* δ_2 and we only have to show that formulae in δ_1 are subformulae of a conclusion or of an hypothesis of δ.

Let us show that $A \odot B$ is a subformula of an hypothesis of Γ_1, which entails the result. What may be the rule R above $A \odot B$?

- if R is $\backslash_e$ or $/_e$, because of the induction hypothesis, and because $A \odot B$ is the conclusion of such a rule, $A \odot B$ is a subformula of Γ_1.
- if R is $\backslash_i$ or $/_i$: this case cannot happen because $A \odot B$ may not be a possible conclusion of those rules.
- if R is $\odot_i$: this case cannot happen: there would exist a 0-extended-redex, *i.e.* a redex and this impossible in a normal proof.
- if R is another $\odot_e$ rule:

$$\cfrac{E \odot F \quad \cfrac{\Gamma_1[E, F] \\ \vdots \, \delta_1 \\ A \odot B}{A \odot B} \, [\odot_e] \quad \cfrac{\Gamma_2[A, B] \\ \vdots \, \delta_2 \\ D}{}}{D} \, [\odot_e]$$

There are two cases:

- If $A \odot B$ can be traced up to an hypothesis of Γ_1, then $A \odot B$ is a subformula of some hypothesis (itself).
- Otherwise, $A \odot B$ is not an hypothesis, but there exists a rule $\odot_i$ above it which generated the formula $A \odot B$. This $\odot_i$ introduction rule is conjoined to the $\odot_e$ rule under discussion, and the proof would contain a k-extended-redex: this case is ruled out.

In any case, $A \odot B$ is subformula of some hypothesis and the subformula property holds for $\odot_e$.

$\square$

Thus in $L_{\odot}$, every proof have a unique normal form which satisfies the subformula property. We observe that unlike [18], rules use are the usual ones for this calculus.

5 Normalisation of proofs of pcIMLL

Now, we present a notion of normal proof with the subformula property, and a normalisation algorithm for proofs of pcIMLL.

As in the previous section about $L_{\odot}$, the normalisation assigns a unique place to the eliminations of non-commutative product, and build sequence of commutative product eliminations. Nevertheless the relative position of each rule in a sequence of elimination rules for commutative product is free, hence not unique unless we accept n-ary rules.

5.1 Some properties of pcIMLL

Property 3 (product elimination rules can swing over any other rule). *Let R be a product elimination rule $\otimes_e$ (resp. $\odot_e$) yielding $\Gamma[\Delta] \vdash C$ from a proof δ_0 of $\Delta \vdash A \odot B$ and a proof of $\Gamma[A, B] \vdash C$ obtained by a rule R' from a proof δ_1 of*

$\Theta[A,B] \vdash X$ (resp. $\Gamma[\langle A;B\rangle] \vdash C$) and from a second proof δ_2 of $\Psi \vdash U$ — if R' is a binary rule.

A proof for the same sequent $\Gamma[\Delta] \vdash C$ can be derived by the following two steps:

1. *apply R between the proof δ_0 of $\Delta \vdash A \odot B$ and the proof δ_1 of $\Theta[A,B] \vdash X$ (resp. $\Theta[\langle A;B\rangle] \vdash X$) as conclusion, yielding $\Theta[\Delta] \vdash X$;*
2. *apply R' to this new proof and to the proof δ_2.*

In other words, provided that the hypotheses A, B that are cancelled by the R product elimination rule are in the same premise of the R' rule, the R product elimination rule can swing over R', as shown in figure 6.

$$\dfrac{\begin{array}{c}\vdots\,\delta_0\\ \Delta \vdash A\otimes B\end{array} \quad \dfrac{\begin{array}{c}\vdots\,\delta_2\\ \Psi \vdash U\end{array} \quad \begin{array}{c}\vdots\,\delta_1\\ \Theta[(A,B)] \vdash X\end{array}}{\Gamma[(A,B)] \vdash C}R'}{\Gamma[\Delta] \vdash C}R \quad \Rightarrow \quad \dfrac{\begin{array}{c}\vdots\,\delta_2\\ \Psi \vdash U\end{array} \quad \dfrac{\begin{array}{c}\vdots\,\delta_0\\ \Delta \vdash A\otimes B\end{array} \quad \begin{array}{c}\vdots\,\delta_1\\ \Theta[(A,B)] \vdash X\end{array}}{\Theta[\Delta] \vdash X}R}{\Gamma[\Delta] \vdash C}R'$$

Figure 6: The product elimination R swings over a rule R' in pcIMLL.

Proof. The proof is similar to the one of property 1. This is a case study according to the rule over the product elimination. This elimination rule can only float up when the hypotheses that must be cancelled are in the same premise and occupy the proper respective position required by the elimination rule.

Let us check that $\otimes_e$ may swing over any other rule R':

- R' is $\backslash_e$:
  - if the hypotheses to be cancelled are in the left premise of $\backslash_e$:

$$\dfrac{\Delta \vdash A\otimes B \quad \dfrac{\Gamma[(A,B)] \vdash D \quad \Phi \vdash D\backslash C}{\langle \Gamma[(A,B)];\Phi\rangle \vdash C}[\backslash_e]}{\langle \Gamma[\Delta];\Phi\rangle \vdash C}[\otimes_e]$$

$$\Rightarrow \quad \dfrac{\dfrac{\Delta \vdash A\otimes B \quad \Gamma[(A,B)] \vdash D}{\Gamma[\Delta] \vdash D}[\otimes_e] \quad \Phi \vdash D\backslash C}{\langle \Gamma[\Delta];\Phi\rangle \vdash C}[\backslash_e]$$

- if the hypotheses to be cancelled are in the right premise of $\backslash_e$:

$$\cfrac{\Delta \vdash A \otimes B \quad \cfrac{\Gamma \vdash D \quad \Phi[(A,B)] \vdash D \backslash C}{\langle \Gamma; \Phi[(A,B)] \rangle \vdash C} \, [\backslash e]}{\langle \Gamma; \Phi[\Delta] \rangle \vdash C} \, [\otimes e]$$

$$\Rightarrow \quad \cfrac{\Gamma \vdash D \quad \cfrac{\Delta \vdash A \otimes B \quad \Phi[(A,B)] \vdash D \backslash C}{\Phi[\Delta] \vdash D \backslash C} \, [\otimes e]}{\langle \Gamma; \Phi[\Delta] \rangle \vdash C} \, [\backslash e]$$

○ R' is $/_e$ — symmetrical to the previous case.

○ R' is $\multimap_e$:

- if the hypotheses to be cancelled are in the left premise of $\multimap_e$:

$$\cfrac{\Delta \vdash A \otimes B \quad \cfrac{\Gamma[(A,B)] \vdash D \quad \Phi \vdash D \multimap C}{(\Gamma[(A,B)], \Phi) \vdash C} \, [\multimap e]}{(\Gamma[\Delta]; \Phi) \vdash C} \, [\otimes e]$$

$$\Rightarrow \quad \cfrac{\cfrac{\Delta \vdash A \otimes B \quad \Gamma[(A,B)] \vdash D}{\Gamma[\Delta] \vdash D} \, [\otimes e] \quad \Phi \vdash D \multimap C}{(\Gamma[\Delta], \Phi) \vdash C} \, [\multimap e]$$

- if the hypotheses to be cancelled are in the right premise of $\multimap_e$:

$$\cfrac{\Delta \vdash A \otimes B \quad \cfrac{\Gamma \vdash D \quad \Phi[(A,B)] \vdash D \multimap C}{(\Gamma, \Phi[(A,B)]) \vdash C} \, [\multimap e]}{(\Gamma, \Phi[\Delta]) \vdash C} \, [\otimes e]$$

$$\Rightarrow \quad \cfrac{\Gamma \vdash D \quad \cfrac{\Delta \vdash A \otimes B \quad \Phi[(A,B)] \vdash D \multimap C}{\Phi[\Delta] \vdash D \multimap C} \, [\otimes e]}{(\Gamma, \Phi[\Delta]) \vdash C} \, [\multimap e]$$

○ R' is $/_i$:

$$\cfrac{\Delta \vdash A \otimes B \quad \cfrac{\langle \Gamma[(A,B)]; D \rangle \vdash C}{\Gamma[(A,B)] \vdash C \mathbin{/} D} \, [/_i]}{\Gamma[\Delta] \vdash C \mathbin{/} D} \, [\otimes_e] \Rightarrow \cfrac{\cfrac{\Delta \vdash A \otimes B \quad \langle \Gamma[(A,B)]; D \rangle \vdash C}{\langle \Gamma[\Delta]; D \rangle \vdash C} \, [\otimes_e]}{\Gamma[\Delta] \vdash C \mathbin{/} D} \, [/_i]$$

○ R' is $\backslash_i$ — symmetrical to the previous case.

○ R' is $\multimap_i$:

$$\cfrac{\Delta \vdash A \otimes B \quad \cfrac{(\Gamma[(A,B)], D) \vdash C}{\Gamma[(A,B)] \vdash D \multimap C} \, [\multimap_i]}{\Gamma[\Delta] \vdash D \multimap C} \, [\otimes_e] \Rightarrow \cfrac{\cfrac{\Delta \vdash A \otimes B \quad (\Gamma[(A,B)], D) \vdash C}{(\Gamma[\Delta], D) \vdash C} \, [\otimes_e]}{\Gamma[\Delta] \vdash D \multimap C} \, [\multimap_i]$$

○ R' is $\otimes_e$ (as R):

• if the hypotheses to be cancelled are in the right premise of $\otimes_e$:

$$\cfrac{\Gamma \vdash A \otimes B \quad \cfrac{\Delta \vdash C \otimes D \quad (\Phi, (A,B), (C,D), \Phi') \vdash E}{(\Phi, (A,B), \Delta, \Phi') \vdash E} \, [\otimes_e]}{(\Phi, \Gamma, \Delta, \Phi') \vdash E} \, [\otimes_e]$$

$$\Rightarrow \cfrac{\Delta \vdash C \otimes D \quad \cfrac{\Gamma \vdash A \otimes B \quad (\Phi, (A,B), (C,D), \Phi') \vdash E}{(\Phi, \Gamma, (C,D), \Phi') \vdash E} \, [\otimes_e]}{(\Phi, \Gamma, \Delta, \Phi') \vdash E} \, [\otimes_e]$$

• if the hypotheses to be cancelled are in the left premise of $\otimes_e$:

$$\cfrac{\Gamma \vdash A \otimes B \quad \cfrac{(\Delta, (A,B), \Delta') \vdash C \otimes D \quad (\Phi, (C,D), \Phi') \vdash E}{(\Phi, \Delta, (A,B), \Delta', \Phi') \vdash E} \, [\otimes_e]}{(\Phi, \Delta, \Gamma, \Delta', \Phi') \vdash E} \, [\otimes_e]$$

$$\Rightarrow \quad \cfrac{\cfrac{\Gamma \vdash A \otimes B \quad (\Delta, (A, B), \Delta') \vdash C \otimes D}{(\Delta, \Gamma, \Delta') \vdash C \otimes D} \ [\otimes_e] \quad (\Phi, (C, D), \Phi') \vdash E}{(\Phi, \Delta, \Gamma, \Delta, \Phi') \vdash E} \ [\otimes_e]$$

○ R' is $\odot_e$:

- if the hypotheses to be cancelled are in the right premise of $\odot_e$:

$$\cfrac{\Gamma \vdash A \otimes B \quad \cfrac{\Delta \vdash C \odot D \quad (\Phi, (A, B), \Psi, \langle C; D \rangle, \Psi', \Phi') \vdash E}{(\Phi, (A, B), \Psi, \Delta, \Psi', \Phi') \vdash E} \ [\odot_e]}{(\Phi, \Gamma, \Psi, \Delta, \Psi', \Phi') \vdash E} \ [\otimes_e]$$

$$\Rightarrow \cfrac{\Delta \vdash C \odot D \quad \cfrac{\Gamma \vdash A \otimes B \quad (\Phi, (A, B), \Psi, \langle C; D \rangle, \Psi', \Phi') \vdash E}{(\Phi, \Gamma, \Psi, \langle C; D \rangle, \Psi', \Phi') \vdash E} \ [\otimes_e]}{(\Phi, \Gamma, \Psi, \Delta, \Psi'\Phi') \vdash E} \ [\odot_e]$$

- if the hypotheses to be cancelled are in the left premise of $\odot_e$:

$$\cfrac{\Gamma \vdash A \otimes B \quad \cfrac{(\Delta, (A, B), \Delta') \vdash C \odot D \quad (\Phi, \Psi, \langle C; D \rangle, \Psi', \Phi') \vdash E}{(\Phi, \Psi, \Delta, (A, B), \Delta', \Psi', \Phi') \vdash E} \ [\odot_e]}{(\Phi, \Psi, \Delta, \Gamma, \Delta', \Psi', \Phi') \vdash E} \ [\otimes_e]$$

$$\Rightarrow$$

$$\cfrac{\cfrac{\Gamma \vdash A \otimes B \quad (\Delta, (A, B), \Delta') \vdash C \odot D}{(\Delta, \Gamma, \Delta') \vdash C \odot D} \ [\otimes_e] \quad (\Phi, \Psi, \langle C; D \rangle, \Psi', \Phi') \vdash E}{(\Phi, \Psi, \Delta, \Gamma, \Delta', \Psi', \Phi') \vdash E} \ [\odot_e]$$

○ R' is $\otimes_i$:

- if the hypotheses to be cancelled are in the left premise of $\otimes_i$:

$$\cfrac{\Gamma \vdash A \otimes B \quad \cfrac{(\Delta, (A, B), \Delta') \vdash C \quad \Phi \vdash D}{(\Delta, (A, B), \Delta', \Phi) \vdash C \otimes D} \ [\otimes_i]}{(\Delta, \Gamma, \Delta', \Phi) \vdash C \otimes D} \ [\otimes_e]$$

$$\Rightarrow \dfrac{\dfrac{\Gamma \vdash A \otimes B \quad (\Delta, (A, B), \Delta') \vdash C}{(\Delta, \Gamma, \Delta') \vdash C} \; [\otimes_e] \qquad \Phi \vdash D}{(\Delta, \Gamma, \Delta', \Phi) \vdash C \otimes D} \; [\otimes_i]$$

- if the hypotheses to be cancelled are in the right premise of $\otimes_i$:

$$\dfrac{\Gamma \vdash A \otimes B \quad \dfrac{\Delta \vdash C \quad (\Phi, (A, B), \Phi') \vdash D}{(\Delta, \Phi, (A, B), \Phi') \vdash C \otimes D} \; [\otimes_i]}{(\Delta, \Phi, \Gamma, \Phi') \vdash C \otimes D} \; [\otimes_e] \quad \text{`}$$

$$\Rightarrow \dfrac{\Delta \vdash C \quad \dfrac{\Gamma \vdash A \otimes B \quad (\Phi, (A, B), \Phi') \vdash D}{(\Phi, \Gamma, \Phi') \vdash D} \; [\otimes_e]}{(\Delta, \Phi, \Gamma, \Phi') \vdash C \otimes D} \; [\otimes_i]$$

○ R' is $\odot_i$:

- if the hypotheses to be cancelled are in the left premise of $\odot_i$:

$$\dfrac{\Gamma \vdash A \otimes B \quad \dfrac{(\Delta, (A, B), \Delta') \vdash C \quad \Phi \vdash D}{\langle (\Delta, (A, B), \Delta'); \Phi \rangle \vdash C \odot D} \; [\odot_i]}{\langle (\Delta, \Gamma, \Delta'); \Phi \rangle \vdash C \odot D} \; [\otimes_e]$$

$$\Rightarrow \dfrac{\dfrac{\Gamma \vdash A \otimes B \quad (\Delta, (A, B), \Delta') \vdash C}{(\Delta, \Gamma, \Delta') \vdash C} \; [\otimes_e] \qquad \Phi \vdash D}{\langle (\Delta, \Gamma, \Delta'); \Phi \rangle \vdash C \odot D} \; [\odot_i]$$

- if the hypotheses to be cancelled are in the right premise of $\odot_i$:

$$\dfrac{\Gamma \vdash A \otimes B \quad \dfrac{\Delta \vdash C \quad (\Phi, (A, B), \Phi') \vdash D}{\langle \Delta; (\Phi, (A, B), \Phi') \rangle \vdash C \odot D} \; [\odot_i]}{\langle \Delta; (\Phi, \Gamma, \Phi') \rangle \vdash C \odot D} \; [\otimes_e]$$

$$\Rightarrow \dfrac{\Delta \vdash C \quad \dfrac{\Gamma \vdash A \otimes B \quad (\Phi, (A, B), \Phi') \vdash D}{(\Phi, \Gamma, \Phi') \vdash D} \; [\otimes_e]}{\langle \Delta; (\Phi, \Gamma, \Phi') \rangle \vdash C \odot D} \; [\odot_i]$$

○ R' is entropy $\sqsubset$:

$$\frac{\dfrac{\Gamma \vdash A \otimes B}{\Gamma' \vdash A \otimes B}\,[\sqsubset] \qquad \Delta[A,B] \vdash D}{\Delta[\Gamma'] \vdash D}\,[\otimes_e] \quad \Rightarrow \quad \frac{\dfrac{\Gamma \vdash A \otimes B \qquad \Delta[A,B] \vdash D}{\Delta[\Gamma] \vdash D}\,[\otimes_e]}{\Delta[\Gamma'] \vdash D}\,[\sqsubset]$$

The elimination of the non-commutative product may swing over any rule. Most cases are easy adaptation of the cases in the proof of the property 1 (and almost similar to the $\otimes_e$ that we exhaustively presented). Let us nevertheless show how $R = \odot_e$ can swings over R' when R' is entropy ($\sqsubset$), since this case did not occur in the proof of property 1.

$$\frac{\dfrac{\Gamma \vdash A \otimes B}{\Gamma' \vdash A \otimes B}\,[\sqsubset] \qquad \Delta[\langle A;B\rangle] \vdash D}{\Delta[\Gamma'] \vdash D}\,[\odot_e] \quad \Rightarrow \quad \frac{\dfrac{\Gamma \vdash A \otimes B \qquad \Delta[\langle A;B\rangle] \vdash D}{\Delta[\Gamma] \vdash D}\,[\odot_e]}{\Delta[\Gamma'] \vdash D}\,[\sqsubset]$$

$$\square$$

The procedure for turning a proof into a normal proof is analogous to the one for $L_\odot$. To do so, we firstly introduce the **redexes** of this calculus and generalise the notion of k-extended redex.

The logic contains seven redexes: one for each implicative connective and two for each product connective (the conjoined introduction could be in the left premise or in the right premise):

○ Redex$_/$: $/_i$ immediately followed by $/_e$.

$$\frac{\dfrac{\langle \Gamma;D\rangle \vdash C}{\Gamma \vdash C/D}\,[/_i] \qquad \begin{array}{c} \vdots\, \delta_1 \\ \Delta \vdash D \end{array}}{\langle \Gamma;\Delta\rangle \vdash C}\,[/_e] \quad \Rightarrow \quad \begin{array}{c} \vdots\, \delta_1 \\ \Delta \vdash D \\ \vdots \\ \langle \Gamma;\Delta\rangle \vdash C \end{array}$$

○ Redex$_\backslash$: $\backslash_i$ immediately followed by $\backslash_e$.

$$\cfrac{\Delta \vdash D \quad \cfrac{\langle D; \Gamma \rangle \vdash C}{\Gamma \vdash D \setminus C}\ [\setminus_i]}{\langle \Delta; \Gamma \rangle \vdash C}\ [\setminus_e] \quad \Rightarrow \quad \begin{array}{c} \vdots\, \delta_1 \\ \Delta \vdash D \\ \vdots \\ \langle \Delta; \Gamma \rangle \vdash C \end{array}$$

o Redex$_{\multimap}$: $\multimap_i$ immediately followed by $\multimap_e$.

$$\cfrac{\Delta \vdash D \quad \cfrac{(D, \Gamma) \vdash C}{\Gamma \vdash D \multimap C}\ [\multimap_i]}{(\Delta, \Gamma) \vdash C}\ [\multimap_e] \quad \Rightarrow \quad \begin{array}{c} \vdots\, \delta_1 \\ \Delta \vdash D \\ \vdots \\ (\Delta, \Gamma) \vdash C \end{array}$$

o Redex$_{\odot}$: $\odot_i$ immediately followed by $\odot_e$ on the left.

$$\cfrac{\cfrac{\Delta_1 \vdash A \quad \Delta_2 \vdash B}{\langle \Delta_1; \Delta_2 \rangle \vdash A \odot B}\ [\odot_i] \quad \Gamma[\langle A; B \rangle] \vdash D}{\Gamma[\langle \Delta_1; \Delta_2 \rangle] \vdash D}\ [\odot_e] \quad \Rightarrow \quad \Gamma[\langle \overset{\displaystyle \vdots\, \delta_1}{A} ; \overset{\displaystyle \vdots\, \delta_2}{B} \rangle] \vdash D$$

o Redex$_{\odot}$: $\odot_i$ immediately followed by $\odot_e$ on the right.

$$\cfrac{\Gamma \vdash A \odot B \quad \cfrac{A \vdash A \quad B \vdash B}{\langle A; B \rangle \vdash A \odot B}\ [\odot_i]}{\Gamma \vdash A \odot B}\ [\odot_e] \Rightarrow \begin{array}{c} \vdots\, \delta_1 \\ \Gamma \vdash A \odot B \end{array}$$

o Redex$_{\otimes}$: $\otimes_i$ immediately followed by $\otimes_e$ on the left.

$$\cfrac{\cfrac{A \quad B}{A \otimes B}\ [\otimes_i] \quad \begin{array}{c} A \quad B \\ \vdots \\ D \end{array}}{D}\ [\otimes_e] \quad \Rightarrow \quad \Gamma[(\overset{\displaystyle \vdots\, \delta_1}{A} , \overset{\displaystyle \vdots\, \delta_2}{B})] \vdash D$$

o Redex$_{\otimes}$: $\otimes_i$ immediately followed by $\otimes_e$ on the right.

$$\cfrac{\begin{array}{c} \vdots \delta_1 \\[-2pt] \vdots \\ \Gamma \vdash A \otimes B \end{array} \qquad \cfrac{A \vdash A \quad B \vdash B}{(A,B) \vdash A \otimes B} \,[\otimes_i]}{\Gamma \vdash A \otimes B} \,[\otimes_e] \qquad \Rightarrow \qquad \begin{array}{c} \vdots \delta_1 \\[-2pt] \vdots \\ \Gamma \vdash A \otimes B \end{array}$$

Here as well, we consider k-extended-redexes, defined as in definition 2:

Definition 4. *Every path of a principal branch $B(S_0)$ of length k — counting every rule, including entropy rules if any — from S_0 to S_n with $|S_0|^r = |S_n|^r$ (i.e. the conclusions of those two sequents are the same), such that $|S_0|$ is the conclusion of an elimination rule R_e and S_n is the conclusion of an introduction rule R_i of the same connective, is called a k-**extended-redex**. Note that 0-extended-redexes are redexes.*

Definition 5. *A proof is said to be in **normal form** whenever it does not contain any k-extended-redexes, $\forall k \in I\!N$.*

5.2   Normalisation of pcIMLL

A proof is in **normal form** if it does not contain any k-extended-redex.

As we did for $L_\odot$, we define the three components of the measure to be used for proving normalisation.

1. Given an *implication elimination* rule R ($\backslash_e$, $/_e$ or $\multimap_e$) with conclusion S_0 the integer $e(R)$ is k if there is a k-extended-redex in $B(S_0)$ (called an implication k-extended-redex over R), and 0 otherwise — the k-rules may include entropy rules.

2. Given a *product elimination* rule R ($\odot_e$ or $\otimes_e$) with S_0 as conclusion, the integer $g(R)$ is k if there is a k-extended-redex in $B(S_0)$ (called a product k-extended-redex over R) and 0 otherwise — the k-rules may include entropy rules.

We introduce the size $|\delta|$ of a proof δ as a triple of integers, with the lexicographic order:

$$|\delta| = \langle r(\delta), e(\delta), g(\delta) \rangle$$

where:

- $r(\delta)$ is simply the number of rules in δ — the number of rules decreases when one reduces a redex, in pcIMLL as in any linear calculus.

- $e(\delta) = \min_{R \in IER(\delta)}(e(R))$ or 0 when there is no implication k-extended-redex, where IER is the set of Implication Elimination Rules.

- $g(\delta) = \min_{R \in PER(\delta)}(g(R))$ or 0 when there is no product k-extended-redex where PER is the set of Product Elimination Rules.

Property 4. *A proof δ is normal if and only if it does not contain 0-extended-redex, and $e(\delta) = g(\delta) = 0$.*

Proof. Let δ a proof of pcIMLL.

○ $e(\delta)$ is the minimal distance between the introduction rule and the elimination rule of an implication k-extended-redex ($\backslash$, $/$ or $\multimap$). If its value is zero, while there is no redex that can be reduced, then there is no implication k-extended-redex.

○ $g(\delta)$ is the distance between the introduction rule and the elimination rule of a product k-extended-redex ($\odot$ or $\otimes$). If its value is zero, while there is no redex that can be reduced, then there is no product k-extended-redex.

Note that the two other redexes could only be 0-extended-redexes. Thus a proof without 0-extended redex and such that $e(\delta) = g(\delta) = 0$ is in normal form. The figure 7 shows example of proofs which are not in normal form. □

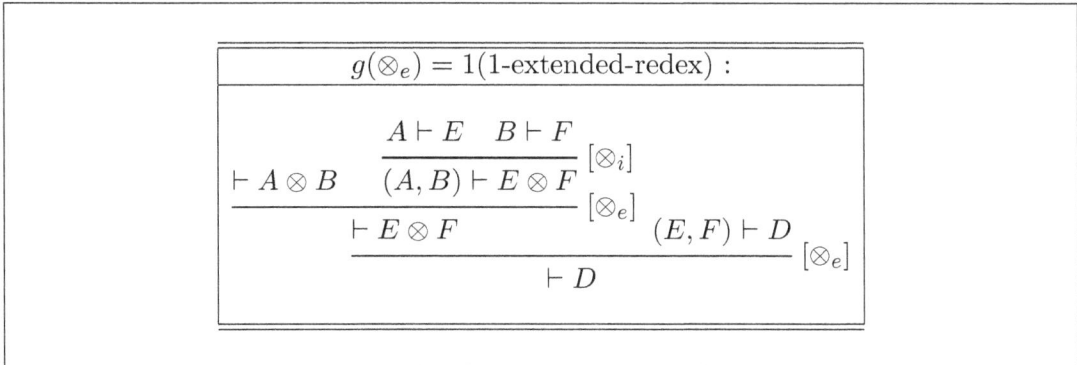

$$g(\otimes_e) = 1(\text{1-extended-redex}) :$$

$$\cfrac{\vdash A \otimes B \quad \cfrac{\cfrac{A \vdash E \quad B \vdash F}{(A, B) \vdash E \otimes F} \, [\otimes_i]}{\vdash E \otimes F} \, [\otimes_e] \quad (E, F) \vdash D}{\vdash D} \, [\otimes_e]$$

Figure 7: Yet another example of a non normal proof in pcIMLL — see figure 4 for other examples

Property 5. *A k-extended-redex $S_0 \cdots S_k$ that includes an implication elimination rule contains a k'-extended-redex, with $k > k'$.*

Proof. Let δ a proof in normal form, and let us consider the principal branch starting with the conclusion.

Any minimal k-extended-redex has the following structure:

$$
\begin{array}{c}
\dfrac{}{X} \; [introduction] \\
\vdots \; \delta_3 \\
\dfrac{U}{U \,/\, A} \; [/_i] \\
\vdots \; \delta_2 \\
\dfrac{U \,/\, A \qquad\qquad\qquad\qquad A}{U} \; [/_e] \\
\vdots \; \delta_1 \\
\dfrac{X}{} \; [elimination]
\end{array}
$$

Then, we define:

- δ_1 as a sequence of implication elimination rules and entropy;

- δ_2 as a sequence of product elimination rules and entropy.

The formula U is the conclusion of the highest implication elimination rule. For this derivation, the number of symbols in U is greater than the number of symbols in X.

Then, still following the principal branch, δ_2 is a sequence of product eliminations and entropy.

The formula $U \,/\, A$ can only result from an introduction rule.

On the example, the only introduction rule that we could structurally use is $/_i$ on the formula A. Because this introduction is in the principal branch, it must be the conjoined rule of the previous introduction. Then, we have a new k'-extended-redex inside the k-extended-redex. k' is the number of rules in δ_2 and because δ_2 is a sub-part of the full proof, $k > k'$. $\qquad\square$

Here is a consequence of the previous property:

Lemma 1. *In the proof δ, a minimal k-extended-redex (whose size is $e(\delta)$) only contains product elimination rules and entropy.*

Proof. If the k-extended-redex is minimal, with the property 5, it does not contain any elimination rule. Moreover, if we do not use elimination rules, the number of symbols in the formula could not decrease. Hence it must be constant in the k-extended-redex. In this case, the rules that we could use are rules whose conclusion is one of the premises. Therefore the sequence of rules may only contain product elimination rules and entropy rules: $\odot_e, \otimes_e$ or $\sqsubset$. □

Property 6. *Product eliminations and entropy rules may swing under implication elimination rules:*

Let R a product elimination rule $\otimes_e$ (resp. $\odot_e$) yielding $\Gamma[\Delta] \vdash C$ from a proof δ_0 of $\Delta \vdash A \otimes B$ and a proof δ_1 of $\Gamma[(A, B)] \vdash C$ (resp. $\Gamma[\langle A; B \rangle] \vdash C$). Assume that the result is merged with a proof δ_2 of $\Theta \vdash U$ via an implication elimination rule R' $\backslash_e$ (resp. $/_e$, $\multimap_e$) yielding $\langle \Theta; \Gamma[\Delta] \rangle \vdash V$ if R' is $\backslash_e$ (resp. $\langle \Gamma[\Delta]; \Theta \rangle \vdash V$ if R', $(\Gamma[\Delta], \Theta) \vdash V$). Figure 8 presents the case where R' is $\backslash_e$.

Then, we can obtain a proof for the same sequent which depends on R' by applying first the rule R' between the proof δ_2 of $\Theta \vdash U$ and the proof δ_1 of $\Gamma[(A, B)] \vdash X$ (resp. $\Gamma[\langle A; B \rangle] \vdash X$) yielding $\langle \Theta; \Gamma[(A, B)] \rangle \vdash V$ (resp. $\langle \Gamma[\langle A; B \rangle]; \Theta \rangle \vdash V$ and $(\Theta, \Gamma[(A, B)]) \vdash V$). Applying the rule R on this new proof, we get the same sequent $\langle \Theta; \Gamma[\Delta] \rangle \vdash V$ (resp. $\langle \Gamma[\Delta]; \Theta \rangle \vdash V$ and $(\Theta, \Gamma[\Delta]) \vdash V$).

Figure 8: The product elimination rule R swings under the rule $\backslash_e$ in pcIMLL.

Proof. Implication eliminations do not modify the order between formulae of the same premise and do modify hypotheses of the premises. Product elimination and entropy rules do not modify the conclusions of the premises but only their hypotheses. Consequently those rule do not interact and may commute. □

Theorem 3. *Every proof δ of pcIMLL has a normal form.*

Proof. We proceed by induction on the size of the proof. Let δ be a proof with $|\delta| = \langle n, e, g \rangle$. By induction, we may assume that every proof δ' of measure $|\delta'| < \langle n, e, g \rangle$ has a normal form.

If δ contains a redex, the reduction of this redex reduces the number of rules in δ, then the resulting proof δ' is such that $n(\delta') < n(\delta)$, hence $|\delta'| < |\delta|$. By induction hypothesis δ' has a normal form, hence δ has one too.

From now on, we can assume without loss of generality that δ has no redex.

If $e(\delta) \neq 0$, then there is an implicative elimination rule S such that S is in a $e(\delta)$-extended-redex. As an $e(\delta)$-extended-redex is minimal, the property 1 implies that it contains only product elimination and entropy rules. Moreover, the property 6 allows to swing S over the rule above it (a product elimination rule or an entropy rules swings under an implication elimination rule). The proof obtained δ' is such that $n(\delta') = n(\delta)$ and $e(\delta') = e(\delta) - 1 < e(\delta)$. By induction hypothesis δ' has a normal form and hence δ has one too.

From now on, we can assume without loss of generality that $e(\delta) = 0$

If $g(\delta) \neq 0$: then there exists a product elimination rule R such that R is in a $g(\delta)$-extended-redex. In this case, δ does not contain any implication extended-redex and $g(R)$ is minimal, hence the extended redex only contains product elimination rule and entropy rules. So R can swing over its left premise (where the rule conjoined $\otimes_i$ is) and over product elimination rule and entropy rule (as property 3 shows). Thus we turned δ into a proof δ' with $n(\delta') = n(\delta)$.

To apply the induction hypothesis we need the size of δ' to be less than the size of δ, and the only thing to check is that no k-extended-redex based on a rule of $IER(\delta)$ appears:

If δ looks like:

○ every principal branch in δ_3 followed by δ_4 does not contain extended-redex because δ_3 is in the left part of $\odot_e$;

○ every principal branch in δ_1 followed by δ_4 may contain extended-redexes;

86

○ every principal branch in δ_2 followed by δ_4 does not contain extended-redexes because δ_2 is in the left part of $\odot_e$ (only for product elimination rules).

The reduction scheme of the redex then gives the new structure of the proof δ':

$$
\begin{array}{c}
A \odot B \\[2ex]
\vdots\ \delta_3 \qquad \vdots\ \delta_1 \\[2ex]
\cfrac{\cfrac{A \odot B \qquad D}{D}\ [\odot_e]}{D}\ [R] \\[2ex]
\vdots\ \delta_4
\end{array}
$$

In this new proof:

○ every principal branch in δ_3 followed by δ_4 does not contain extended-redexes because δ_3 is in the left part of $\odot_e$;

○ every principal branch in δ_1 followed by δ_4 does not contain new extended-redexes, and the measure of these extended-redexes is decremented by 1;

○ every principal branch in δ_2 followed by δ_4 does not contain extended-redexes because δ_2 is in the left part of R (which is necessary a product elimination rule).

The proof does not contain any new k-extended-redex ; in particular the proof does not contain any new implication k-extended-redex. Then, we have $e(\delta') = e(\delta)$ and $g(\delta') = g(\delta) - 1$. Thus $|\delta'| < |\delta|$ and by induction δ' has a normal form, and therefore δ also has one.

If none of the previous transformations applies we have $e(\delta) = g(\delta) = 0$, and therefore, because of property 4, δ is in normal form.

□

Figure 9 is the normal form of the example from figure 7, obtained by following the procedure described in the proof above.

Now, let us establish that proofs in normal form enjoy the subformula property.

$$\frac{g(\otimes_e) = 0(\text{no more } k\text{-extended-redex}) :}{\dfrac{\vdash A \otimes B \qquad \dfrac{\dfrac{A \vdash E \qquad B \vdash F}{(A,B) \vdash E \otimes F}\,[\otimes_i] \qquad (E,F) \vdash D}{(A,B) \vdash D}\,[\otimes_e]}{\vdash D}\,[\otimes_e]}$$

Figure 9: Normal proof for the proof in figure 7

5.3   Subformula property for pcIMLL

Theorem 4. *The subformula property holds for pcIMLL: in a normal proof δ of a sequent $\Gamma \vdash C$, every formula of a sequent is a subformula of some hypothesis (Γ) or of the conclusion (C).*

Proof. We proceed by induction on the number of rules in the normal proof. Once again, we prove a stronger property:

1. every formula in a normal proof is subformula of some hypotheses or of the conclusion of the proof;

2. if the last rule used is an implicative elimination $\backslash_e$, $/_e$ or $\multimap_e$ every subformula is a subformula of some hypothesis.

Axioms enjoy the subformula property.

When the last rule is an entropy rule, the subformula property holds simply because of the induction hypothesis. Indeed, the formulae of a sequent are preserved under entropy rule which only affects the order on the formulae.

Let us call R^* the last rule of the proof.

1. R^* is $\backslash_e$:

$$\frac{\begin{array}{cc} \Delta_1 & \Gamma_2 \\ \vdots\, \delta_1 & \vdots\, \delta_2 \\ \Delta \vdash C & \Gamma \vdash C \backslash D \end{array}}{\langle \Delta; \Gamma \rangle \vdash D}\,[\backslash_e]$$

88

By induction hypothesis, every formula in δ_1 is a subformula of some hypothesis in Δ_1 or of the conclusion C. In addition every formula in δ_2 is a subformula of Γ_2 or of the conclusion $C \backslash D$. However C is a subformula of $C \backslash D$ and D too. Let us show that $C \backslash D$ is subformula of some hypothesis in δ_2.

Let us look at the rule R' that yields $C \backslash D$:

- if R' is $\backslash_i$: this may not happen because it would be a 0-extended-redex while the proof is in normal form.

- if R' is $/_i$, $\multimap_i$, $\otimes_i$ or $\odot_i$: these cases are structurally impossible because they can not produce $C \backslash D$.

- if R' is $\backslash_e$, $/_e$ or $\multimap_e$: we use the induction hypothesis and $C \backslash D$ is a subformula of Γ_2.

- if R' is $\otimes_e$, $\odot_e$ or entropy: they preserve the conclusion, thus we have to investigate what the rule above can be:

  If it is one of the previous rule, we use the same argument. Else, the proof is a finite sequence of $\otimes_e$, $\odot_e$ and entropy. Those rules preserve the conclusion and therefore $C \backslash D$ is one of the hypothesis in Γ_2.

2. R^* is $/_e$ or $\multimap_e$: similar to $\backslash_e$ above.

3. R^* is $\multimap_i$, let δ be the following proof:

$$
\frac{\begin{array}{c} \Gamma \\ \vdots\, \delta_1 \\ \langle \Gamma ; C \rangle \vdash D \end{array}}{\Gamma \vdash C \multimap D}[\multimap_i]
$$

By induction hypothesis, every formula in δ_1 is a subformula of some hypothesis in Γ or of the conclusion D. The formula D is a subformula of $D \multimap C$, hence every formula of δ is a subformula of some hypothesis Γ or of the conclusion $D \multimap C$. The property holds for δ.

4. R^* is $\backslash_i$ or $/_i$ — similar to the previous case.

5. R^* is $\otimes_i$: let δ be the following proof:

$$
\frac{\begin{array}{cc} \begin{array}{c} \Delta_1 \\ \vdots\, \delta_1 \\ \Delta \vdash C \end{array} & \begin{array}{c} \Gamma_2 \\ \vdots\, \delta_2 \\ \Gamma \vdash D \end{array} \end{array}}{(\Delta, \Gamma) \vdash C \otimes D}[\otimes_i]
$$

- every formula in δ_1 is a subformula of some hypothesis in Δ_1 or of the conclusion C.

- every formula in δ_2 is a subformula of some hypothesis in Γ_2 or of the conclusion D.

- however C and D are themselves subformulae of $C \otimes D$, then in Γ, every formula is a subformula of some hypotheses in Δ and Γ or of the conclusion $C \otimes D$.

6. R^* is $\odot_i$: similar to the previous case.

7. R^* is $\otimes_e$:

$$
\frac{
\begin{array}{cc}
\begin{array}{c} \Delta_1 \\ \vdots\ \delta_1 \\ \Delta \vdash A \otimes B \end{array}
&
\begin{array}{c} \Gamma_2 \\ \vdots\ \delta_2 \\ \Gamma[A, B] \vdash D \end{array}
\end{array}
}{
\Gamma[\Delta] \vdash D
} [\otimes_e]
$$

- every formula of δ_1 is a subformula of some hypothesis in Δ_1 or of the conclusion $A \otimes B$.

- every formula of δ_2 is a subformula of some hypothesis in Γ_2 or of the conclusion D.

- moreover, D is the conclusion of δ. Thus, every formula of δ_2 is a subformula of some hypothesis in Γ_2 or of the conclusion of the proof δ: D.

In order to prove that the property holds for the other part of the proof, we must prove that $A \otimes B$ is a subformula of some hypothesis in δ_1. Let us look at the rule R' above:

o if R' is $\backslash_e$, $/_e$ or $\multimap_e$, using the induction hypothesis $A \otimes B$ is subformula of hypotheses Δ_1.

o if R' is $\otimes_i$: this case is impossible because there cannot be any 0-extended-redex in a normal proof.

o if R' is $\backslash_i$, $/_i$, $\multimap_i$ or $\odot_i$: these cases are structurally impossible because these rules cannot produce $A \otimes B$.

o if R' is $\otimes_e$, $\odot_e$ or entropy which preserve the conclusion of the proof, let us analyse the rule above:

- either it is one of the previous rules, thus, using the same arguments we conclude.
- either, given that a proof only contains a finite number of rules, the sequence of such rules is finite. Givven that it contains only $\otimes_e$, $\odot_e$ and entropy rules, thus the formula is a hypothesis in Δ_1.

In every possible case, $A \otimes B$ or $A \odot B$ is subformula of hypotheses.

8. R^* is $\odot_i$: similar to the previous case.

In pcIMLL, all proofs have a normal form which enjoys the subformula property. $\square$

6 Decidability

An immediate but interesting consequence of normalisation with a subformula property is the following:

Theorem 5. *The provability of a sequent in pcIMLL is decidable, and in $L_\odot$ as well.*

Proof. Because of normalisation, one only has to look for normal proofs. Given that normal proofs enjoy the subformula property it is enough to try the *finite* number of rules that are possible. There are finitely many rules, and each of them may only lead to try to prove a finite number of sequents because of the subformula property. By considering principal branches, premises of these rules are sequents that have less connectives. Therefore an easy induction shows that the calculus is decidable. For more details, see the proof of decidability for product free Lambek calculus based on natural deduction in [17]. $\square$

7 Conclusion

With concurrency and linguistics motivations, we defined pcIMLL in natural deduction and proved normalisation. For Lambek calculus with product, a subcalculus of pcIMLL, we also characterised the unique normal form.

As a perspective, we look forward a proof net syntax for pcIMLL. This would also allow to easily compute lambda terms (that are semantic reading in linguistic applications). Although related systems do have proof nets (MLL, Lambek calculus, NL of Abrusci and Ruet) there is not yet any proof net calculus for pcIMLL. The present work on natural deduction can be viewed as a first step in this direction.

We avoided tricky details and discussions about the uniqueness of the normal form for pcIMLL. Let us say it can be achieved if one consider as equivalent proofs that only differ because of the relative order of several commutative product elimination rules in a sequence of product eliminations that are just below the rule which gathers the cancelled hypotheses.

With respect to computational linguistic application, we look forward a simpler translation from pcIMLL formulae to arrow types on e and t and thus from parse structures that are pcIMLL deduction to intuitionistic deductions, which are semantic readings. This is related to the interpretation of noun phrase and generalised quantifiers as the combination of the categories k (case) and d (entities).

References

[1] Samson Abramsky. Computational interpretations of linear logic. *Theoretical Computer Science*, 111:3–57, 1993.

[2] V. Michele Abrusci. Phase semantics and sequent calculus for pure noncommutative classical linear propositional logic. *The Journal of Symbolic Logic*, 56(4):1403–1451, December 1991.

[3] V. Michele Abrusci and Paul Ruet. Non-commutative logic I: The multiplicative fragment. *Annals of pure and applied logic*, 101(1):29–64, 1999.

[4] Maxime Amblard. *Calcul de représentations sémantiques et suntaxe générative: les grammaires minimalistes catégorielles*. PhD thesis, université de Bordeaux 1, 2007.

[5] Maxime Amblard. Encoding Phases using Commutativity and Non-commutativity in a Logical Framework. In Sylvain Pogodalla and Jean-Philippe Prost, editors, *Logical Aspect of Computational Linguistic*, volume 6736 of *Lecture Notes in Computer Science*, pages 1–16, Montpellier, France, June 2011. Springer.

[6] Maxime Amblard, Alain Lecomte, and Christian Retoré. Categorial minimalist grammars: From generative grammar to logical form. *Linguistic Analysis*, 36(1–4):273–306, 2010. Festschrift on the occasion of Jim Lambek's 85th birthday.

[7] Denis Bechet, Philippe de Groote, and Christian Retoré. A complete axiomatisation of the inclusion of series-parallel partial orders. In H. Comon, editor, *Rewriting Techniques and Applications, RTA'97*, volume 1232 of *LNCS*, pages 230–240. Springer Verlag, 1997.

[8] Philippe de Groote. Partially commutative linear logic: sequent calculus and phase semantics. In Vito Michele Abrusci and Claudia Casadio, editors, *Third Roma Workshop: Proofs and Linguistics Categories – Applications of Logic to the analysis and implementation of Natural Language*, pages 199–208. Bologna:CLUEB, 1996.

[9] Samuel Epstein and Robert Berwick. On the convergence of 'minimalist' syntax and categorial grammar. In A. Nijholt, G. Scollo, and R. Steetkamp, editors, *Algebraic Methods in Language Processing*. Universiteit Twente, 1995.

[10] Jean-Yves Girard. Linear logic. *Theoretical Computer Science*, 50(1):1–102, 1987.

[11] Jean-Yves Girard, Yves Lafont, and Paul Taylor. *Proofs and Types*. Number 7 in Cambridge Tracts in Theoretical Computer Science. Cambridge University Press, 1988.

[12] Alessio Guglielmi. A system of interaction and structure. *ACM Trans. Comput. Logic*, 8(1), January 2007.

[13] J.. S. Hodas and D. Miller. Logic programming in a fragment of intuitionistic linear logic. *Information and computation*, pages 327–365, 1994.

[14] Joachim Lambek. The mathematics of sentence structure. *American mathematical monthly*, pages 154–170, 1958.

[15] Alain Lecomte and Christian Retoré. Extending Lambek grammars: a logical account of minimalist grammars. In *Proceedings of the 39th Annual Meeting of the Association for Computational Linguistics, ACL 2001*, pages 354–361, Toulouse, July 2001. ACL.

[16] Marcel Masseron, Christophe Tollu, and Jacqueline Vauzeilles. Generating plans in linear logic: I. actions as proofs. *Theoretical Computer Science*, 113:349–370, 1993.

[17] Richard Moot and Christian Retoré. *The logic of categorial grammars: a deductive account of natural language syntax and semantics*, volume 6850 of *LNCS*. Springer, 2012. http://www.springer.com/computer/theoretical+computer+science/book/978-3-642-31554-1.

[18] Sara Negri. A normalizing system of natural deduction for intuitionistic linear logic. *Archive for Mathematical Logic*, 2002.

[19] Christian Retoré. *Réseaux et Séquents Ordonnés*. Thèse de Doctorat, spécialité Mathématiques, Université Paris 7, février 1993.

[20] Christian Retoré. Pomset logic: a non-commutative extension of classical linear logic. In Philippe de Groote and James Roger Hindley, editors, *Typed Lambda Calculus and Applications, TLCA'97*, volume 1210 of *LNCS*, pages 300–318, 1997.

[21] Christian Retoré. A description of the non-sequential execution of petri nets in partially commutative linear logic. In Jan van Eijck, Vincent van Oostrom, and Albert Visser, editors, *Logic Colloquium 99*, Lecture Notes in Logic, pages 152–181. ASL and A. K. Peters, 2004.

[22] Christian Retoré and Edward Stabler. Generative grammar in resource logics. *Research on Language and Computation*, 2(1):3–25, 2004. Introductory survey for the special issue on *Resource logics and minimalist grammars*.

[23] Paul Ruet. *Logique non-commutative et programmation concurrente*. Thèse de doctorat, spécialité logique et fondements de l'informatique, Université Paris 7, 1997.

[24] Edward Stabler. Derivational minimalism. In Christian Retoré, editor, *Logical Aspects of Computational Linguistics, LACL'96*, volume 1328 of *LNCS/LNAI*, pages 68–95. Springer-Verlag, 1997.

Received January 2014

Tools for Conviviality in Multi-Context Systems

Antonis Bikakis

Department of Information Studies, University College London, UK
a.bikakis@ucl.ac.uk

Patrice Caire and Yves Le Traon

University of Luxembourg, Interdisciplinary Center for Security, Reliability and Trust, Luxembourg
{patrice.caire,yves.letraon}@uni.lu

Abstract

A common feature of many distributed systems, including web social networks, peer-to-peer systems and Ambient Intelligence systems, is cooperation in terms of information exchange among heterogeneous entities. In order to facilitate the exchange of information, we first need ways to evaluate it. The concept of conviviality was recently proposed for modeling and measuring cooperation among agents in multiagent systems. In this paper, we introduce conviviality as a property of Multi-Context Systems (MCS). We first present how to use conviviality to model and evaluate interactions among different contexts, which represent heterogeneous entities in a distributed system. Then, as one cause of logical conflicts in MCS is due to the exchange of information between mutually inconsistent contexts, we show how inconsistency can be resolved using the conviviality property. We illustrate our work with an example from web social networks.

1 Introduction

Multi-Context Systems (MCS) [19, 18, 9] are logical formalizations of distributed context theories connected through a set of bridge rules, which enable information flow between contexts. A *context* can be thought of as a logical theory - a set of axioms and inference rules - that models local knowledge. Intuitively, MCS can

---

Thanks to the National Research Fund, Luxembourg (CO11/IS/1239572) CoPAInS project.

be used as a representation model for any information system that involves distributed, heterogeneous knowledge agents such as peer-to-peer systems, distributed ontologies (e.g., Linked Open Data) or Ambient Intelligence systems. In fact, several applications have already been developed on top of MCS or other logic-based context formalizations including (a) the CYC common sense knowledge base [23]; (b) contextualized ontology languages, such as Distributed Description Logics [5] and C-OWL [6]; (c) context-based agent architectures [24, 25]; and (d) distributed reasoning algorithms for Mobile Social Networks [1] and Ambient Intelligence systems [2].

The individual entities that such systems consist of cooperate by sharing information. By reasoning with the information they import they are able to derive new knowledge. These features are enabled by the notions of *contexts*, *bridge rules* and *contextual reasoning* used in MCS. But, how can we then evaluate the ways in which a system enables this cooperation? How can we characterise a MCS based on the opportunities for information exchange that it provides to its contexts? To answer such questions, we build on previous work on modeling *conviviality* in a version of MCS called Contextual Defeasible Logic [12]. Here we extend these results for the general MCS model, and introduce measures for information dependencies based again on the notion of conviviality.

Defined by Illich as "individual freedom realized in personal interdependence" [21], conviviality was introduced as a social science concept for multiagent systems to highlight soft qualitative requirements like user friendliness of systems. Multiagent systems technology can be used to realize tools for conviviality when "freedom" is interpreted as choice [10]. Tools for conviviality are concerned in particular with dynamic aspects of conviviality, such as the emergence of conviviality from the sharing of properties or behaviors whereby each member's perception is that their personal needs are taken care of.

Conviviality is measured by counting the possible ways to cooperate, indicating degree of choice or freedom to engage in coalitions [11]. The authors' coalitional theory is based on dependence networks [13, 28], labeled directed graphs where nodes represent agents, (thus the graph represents a social network), and each labeled edge represents that the former agent depends on the latter one to achieve some goal (represented by the label).

The focus on dependence networks and specifically on their cycles is a reasonable way of formalizing conviviality as something related to the freedom of choice of individuals plus the subsidiary relations – interdependence for task achievement – among fellow members of a social system. In distributed information systems, individual freedom is linked to an agent's choice to keep personal knowledge and beliefs at the local level, while interdependence is understood as reciprocity, i.e.

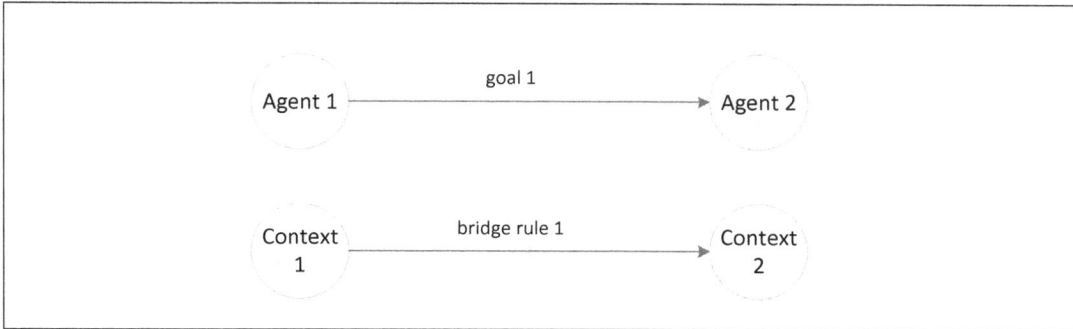

Figure 1: The dependence network parallelism of contexts as agents, and bridge rules as goals. A labeled arrow, from *Agent1* to *Agent2* means that the former depends on the latter to achieve its *goal1*.

cooperation. Participating human and artificial entities depend on each other to increase their local knowledge.

In this paper, we draw a parallel between, on the one hand an agent and a context, and on the other hand a goal and a bridge rule. More specifically, we use a context to encode an agent's knowledge in some logic language, and a bridge rule to describe how an agent achieves its goal, namely to acquire and combine knowledge from other agents in order to deduce new knowledge, as illustrated in Figure 1. Therefore, the dependency from an agent A_1 towards a distinct agent A_2 to fulfill its goal g_1 corresponds to the context C_1 depending on a distinct context C_2 to acquire knowledge through the exchange of information described in a rule r_1. Furthermore, evaluating this exchange would allow to reason about the system with respect to how it can be reconfigured to enable more cooperation among contexts and thereby more information sharing, opportunities to collaborate and possibility to choose among them. Particularly, considering the potential applications of MCS, and the tools for conviviality described above, we formulate our main research question as follows:

> *How to evaluate and improve the exchange of information in systems modeled as MCS using conviviality modeling and measures?*

Our main research question breaks into the following questions:

1. How to define and model conviviality for Multi-Context Systems?

2. How to measure the conviviality of Multi-Context Systems?

3. How to use conviviality as a property of Multi-Context Systems?

In this paper we address these questions by proposing the following:

1. A formal model for representing *information dependencies* in MCS based on dependence networks,

2. Conviviality measures for MCS and

3. A potential application of these tools for the problem of inconsistency resolution in MCS.

So far, most approaches for inconsistency resolution in MCS have been based on the *invalidation* or *unconditional application* of a subset of the bridge rules that cause inconsistency. They differ in the preference criterion that is applied for selecting among the candidate solutions. In this work, we propose to use the conviviality of the system as a preference criterion, based on the idea that removing (or applying unconditionally) a bridge rule affects the information dependency between the connected contexts, and, as a result, the conviviality of the system. We suggest that the optimal solution is the one that minimally reduces conviviality.

The paper is structured as follows: Section 2 describes our running example from the social web application domain. Section 3 presents formal definitions for MCS, as these were originally proposed in [9]. Section 4 introduces a model and measures for conviviality in MCS. Section 5 proposes a potential use of conviviality as a property of MCS for the problem of inconsistency resolution. Section 6, presents and compares related works. The last section summarizes and provides insights on our future works.

2 Running Example

In order to demonstrate the exchange of information among heterogeneous agents, we use an example from the domain of social networks, namely a social web application, and highlight the requirements and challenges with respect to knowledge sharing and collective decision making.

A typical challenge for students is to find how to organize their references. They need to record their readings and have quick and easy access to research articles. Therefore articles need to be classified in a way that is tailored to their studies. Furthermore, if more students contribute to this classification, more articles will be available to the whole group for citations.

Jane, Bob and Charlie are members of uni.scholar.space. They use software agents ($A1$, $A2$ and $A3$ respectively) to connect to the network in order to share information and classify research articles that they find online. The three agents are heterogeneous with respect to their capabilities, the knowledge that they encode, and the *logic* with which they represent and reason with the available knowledge.

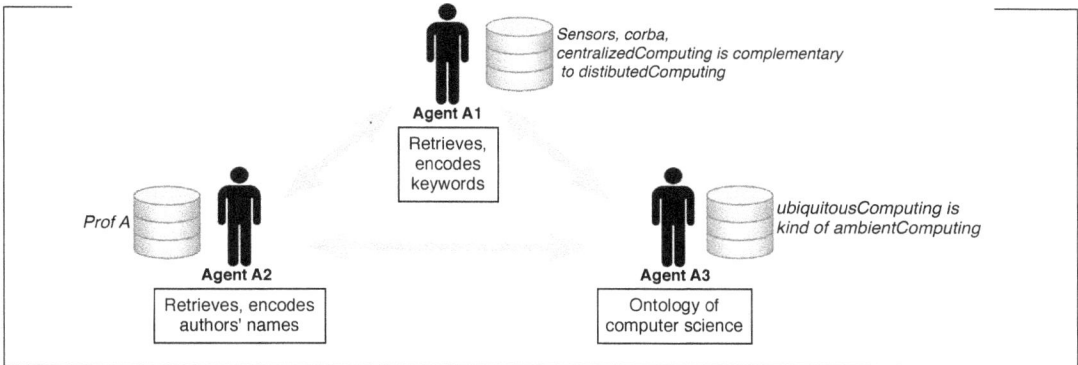

Figure 2: Students' social network for information exchange and collective classification of articles; ideal case where all agents cooperate with each other.

$A1$ retrieves the keywords of articles and encodes this information as well as Jane's research knowledge in propositional logic. $A2$ uses propositional logic as well to encode information about the authors of the articles and Bob's research knowledge. Finally, $A3$ contains an ontology about Computer Science written by Charlie in a basic description logic.

The system enables the heterogeneous agents to exchange their local knowledge, and take together decisions about the classification of an article by exploiting as much as possible from the available information. Figure 2 illustrates the ideal case where all agents are enabled to exchange information with each other and take a collective decision about the classification of an article.

To make the example more concrete we consider a specific article for which the three agents have retrieved the following metadata: the article has two keywords, *sensors* and *corba*, and is written by *Prof.A*. Moreover, according to $A1$, centralized computing and distributed computing are two complementary concepts; and according to $A3$ ubiquitous computing is a form of ambient computing. In order to be able to exchange information, the three users have identified the following mappings between the concepts that they use: for Jane ($A1$) the term *middleware* used by Bob ($A2$) implies centralized computing, while the term *ambient computing* used by Charlie ($A3$) implies distributed computing. Bob knows that *corba* stands for Common Object Request Broker Architecture, and is a type of middleware. Finally, for Charlie, articles that are written by Prof. B and are about *sensors* are relevant to ubiquitous computing.

In the following sections we show how information exchange between heterogeneous agents such as the ones in our running example is enabled by MCS; and how we can evaluate cooperation between agents in terms of opportunities for information exchange by using conviviality as a property of MCS.

3 Multi-Context Systems

We use here the definition of heterogeneous nonmonotonic MCS given in [7]. The idea behind heterogeneous MCSs is to allow different logics to be used in different contexts, and to model information flow among contexts via bridge rules. According to [7], a MCS M is a set of contexts, each composed of a knowledge base with an underlying logic, and a set of bridge rules. A logic $L = (\mathbf{KB}_L, \mathbf{BS}_L, \mathbf{ACC}_L)$ consists of the following components:

- $\mathbf{KB}_L$ is the set of well-formed knowledge bases of L. Each element of $\mathbf{KB}_L$ is a set of formulae.

- $\mathbf{BS}_L$ is the set of possible belief sets, where the elements of a belief set is a set of formulae.

- $\mathbf{ACC}_L$: $\mathbf{KB}_L \to 2^{\mathbf{BS}_L}$ is a function describing the semantics of the logic by assigning to each knowledge base a set of acceptable belief sets.

As shown in [7], this definition captures the semantics of many different logics both monotonic, e.g. propositional logic, description logics and modal logics, and non-monotonic, e.g. default Logic, circumscription, defeasible logic and logic programs under the answer set semantics.

A *bridge rule* refers in its body to other contexts and can thus add information to a context based on what is believed or disbelieved in other contexts. Bridge rules are added to those contexts to which they potentially add new information. Let $L = (L_1, \ldots, L_n)$ be a sequence of logics. An L_k-bridge rule r over L, $1 \leq k \leq n$, is of the form

$$r = (k : s) \leftarrow (c_1 : p_1), \ldots, (c_j : p_j),$$
$$\mathbf{not}(c_{j+1} : p_{j+1}), \ldots, \mathbf{not}(c_m : p_m). \tag{1}$$

where c_i, $1 \leq c_i \leq n$, refers to a context in M, p_i is an element of some belief set of L_{c_i}, and k refers to the context receiving information s. We denote by $h_b(r)$ the belief formula s in the head of r. By $br_M = \bigcup_{i=1}^{n} br_i$ we denote the set of bridge rules in M.

A *MCS* $M = (C_1, \ldots, C_n)$ is a set of contexts $C_i = (L_i, kb_i, br_i)$, $1 \leq i \leq n$, where $L_i = (\mathbf{KB}_i, \mathbf{BS}_i, \mathbf{ACC}_i)$ is a logic, $kb_i \in \mathbf{KB}_i$ a knowledge base, and br_i a set of L_i-bridge rules over $(L_1, \ldots, L_n)$. For each $H \subseteq \{h_b(r)|r \in br_i\}$ it holds that $kb_i \cup H \in \mathbf{KB}_{L_i}$, meaning that bridge rule heads are compatible with knowledge bases.

Example 3.1. Agents $A1$, $A2$ and $A3$ of our running example can be modeled as contexts C_1, C_2 and C_3 respectively in a MCS $M = \{C_1, C_2, C_3\}$. The knowledge bases of the three contexts are:

$$kb_1 = \{sensors, corba, centralizedComputing \leftrightarrow \neg distributedComputing\}$$
$$kb_2 = \{profA\}$$
$$kb_3 = \{ubiquitousComputing \subseteq ambientComputing\}$$

The bridge rules that the three agents use to exchange information and collectively decide about the classification of the article are as follows:

$$r_1 = (1 : centralizedComputing) \leftarrow (2 : middleware)$$
$$r_2 = (1 : distributedComputing) \leftarrow (3 : ambientComputing)$$
$$r_3 = (2 : middleware) \leftarrow (1 : corba)$$
$$r_4 = (3 : ubiquitousComputing) \leftarrow (1 : sensors), (2 : profB)$$

A belief state of a MCS is the set of the belief sets of its contexts. Formally, a *belief state* of $M = (C_1, \ldots, C_n)$ is a sequence $S = (S_1, \ldots, S_n)$ such that $S_i \in \mathbf{BS}_i$. Intuitively, S is derived from the knowledge of each context and the information conveyed through applicable bridge rules. A bridge rule of form (1) is applicable in a belief state S iff for $1 \le i \le j$: $p_i \in S_{c_i}$ and for $j < l \le m$: $p_l \notin S_{c_l}$.

Equilibrium semantics selects certain belief states of a MCS as acceptable. Intuitively, for a MCS $M = (C_1, \ldots, C_n)$, an equilibrium is a belief state $S = (S_1, \ldots, S_n)$ where each context C_i respects all bridge rules that are applicable in S and accepts S_i. Formally, S is an equilibrium of M, iff for $1 \le i \le n$,

$$S_i \in \mathbf{ACC}_i(kb_i \cup \{h_b(r)|r \in br_i \text{ applicable in } S\}).$$

Example 3.2. In our running example, $S = (S_1, S_2, S_3)$ is the only equilibrium of the system:

$$S = (\{sensors, corba, centralizedComputing\}, \{profA, middleware\}, \emptyset).$$

S_3 is an empty set, since r_4, which is the only bridge rule in C_3, is not applicable in S, because $profB \notin S_2$.

4 The Conviviality Property in MCS

We recall from Section 1, that dependence networks have been proposed as a model for representing social dependencies among the agents of a multiagent system. They have also been used as the underlying model for formalizing and measuring conviviality in such systems. In this section, we describe how dependence networks can be used to model information dependencies among the contexts of a MCS and how conviviality measures can then be applied to MCS.

Our approach is based on the following ideas. First, cooperation in MCS can be understood as information sharing among its contexts. Second, this cooperation is enabled by the bridge rules of the system. Hence, finally, bridge rules actually represent information dependencies among contexts. On one hand, the more bridges between the contexts, the more possibilities for cooperation and information exchange. On the other hand, no bridge rules would mean that the different contexts represent autonomous systems, which do not share their local knowledge.

4.1 Model

Conviviality can be modeled by the reciprocity-based coalitions, or group of agents, that may be formed [11]. Some coalitions, however, provide more opportunities for their participants to cooperate than others, being thereby more convivial. Dependence networks are used to represent the interdependencies among the participants of the coalitions. Abstracting from tasks and plans that agents may have to achieve their goals, a dependence network for a multiagent system is defined [11] as follows:

Definition 4.1 (Dependence network). *A dependence network (DN) is a tuple* $\langle A, G, dep, \geq \rangle$ *where: A is a set of agents, G is a set of goals, $dep : A \times A \rightarrow 2^G$ is a function that relates with each pair of agents, the sets of goals on which the first agent depends on the second, and $\geq : A \rightarrow 2^G \times 2^G$ is for each agent a total pre-order on sets of goals occurring in its dependencies: $G_1 >_{(a)} G_2$.*

To capture the notions of *context* and *bridge rule*, we build on Definition 4.1 and introduce a new definition, Definition 4.2, for a dependence network that corresponds to a MCS, as follows:

Definition 4.2 (Dependence network for MCS). *A dependence network corresponding to a MCS M, denoted as $DN(M)$, is a tuple $\langle C, br_M, dep, \geq \rangle$ where: C is the set of contexts in M; br_M is the set of bridge rules in M; $dep : C \times C \rightarrow 2^{br_M}$ is a function that is constructed as follows: for each bridge rule r (in the form of (1)) in br_M add the following dependencies: $dep(k, c_i) = \{r\}$ where k is the context appearing in the head of r and c_i stands for each distinct context appearing in the*

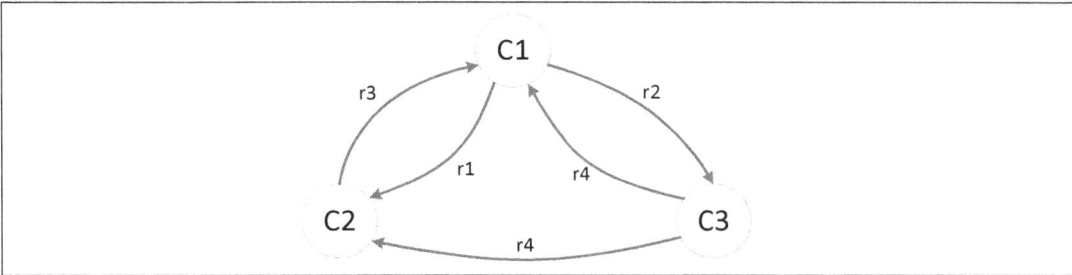

Figure 3: The dependence network $DN(M)$ of MCS M of the running example, in which a specific article is under examination.

body of r; and $\geq: C \rightarrow 2^{br_M} \times 2^{br_M}$ is for each context a total pre-order on sets of its bridge rules.

In other words, a bridge rule r creates one dependency between context k, which appears in the head of r, and each of contexts c_i that appear in the body of r. The intuition behind this is that k depends on the information it receives from each c_i to achieve its goal, which is to apply r in order to infer s.

We should also note here that the total preorder that each context defines on the sets of bridge rules may reflect the local preferences of a context, e.g., in the way that these are defined and used in Contextual Defeasible Logic [2]. For sake of simplicity, we do not use this feature in the conviviality model that we describe in this paper. However, it is among our plans to integrate it in future extensions of this work. To graphically represent dependence networks, we use nodes for contexts and labeled arrows for dependencies among the contexts that the arrows connect. An arrow from context a to context b, labeled as r, means that a depends on b to apply rule r.

Example 4.3. In our running example, the dependence network that corresponds to MCS M is $DN(M) = \langle C, br_M, dep, \geq \rangle$ where:

- $C = \{C_1, C_2, C_3\}$ is the set of contexts in M

- $br_M = \{r_1, r_2, r_3, r_4\}$ is the set of bridge rules in M

- The dependencies, as per Definition 4.2, are the following:
  $dep(C_1, C_3) = \{r_2\}$, $dep(C_3, C_1) = \{r_4\}$, $dep(C_1, C_2) = \{r_1\}$,
  $dep(C_2, C_1) = \{r_3\}$, $dep(C_3, C_2) = \{r_4\}$;

The graphical representation of the dependence network is illustrated in Figure 3. The figure should be read as follows: each node corresponds to one of the contexts in M. Dependencies are derived from the four bridge rules of M. For example, there are two dependencies labeled by r_4: each of them connects C_3, which appears in the head of r_4, to each distinct context appearing in the body of r_4, namely C_1 and C_2 respectively. This actually means that to apply rule r_4 in order to prove that the paper under examination is about ubiquitous computing, C_3 depends on information about the keywords of the paper that it imports from C_1 and information about the authors of the paper that it imports from C_2.

To evaluate MCS in terms of the information exchange we introduce appropriate measures in the next section.

4.2   Measures

Conviviality measures were introduced to compare the conviviality of multiagent systems [11], for example before and after making a change such as adding a new norm or policy. Furthermore, to evaluate conviviality in a more precise way, the authors introduced formal conviviality measures for dependence networks using a coalitional game theoretic framework. Based on Illich's definition of conviviality as "individual freedom realized in personal interdependency", the notions of interdependency and choice, if freedom is interpreted as choice, are emphasized. Such measures provide insights into the type of attributes that may be measured in a convivial system and thus evaluate the quality of the system from this point of view. The conviviality measures we present here reflect the following hypotheses:

H1 *The cycles identified in a dependence network are considered as coalitions, i.e., grouping of contexts. Such coalitions are used to evaluate conviviality in the network. Cycles are the smallest graph topology expressing interdependence, thereby conviviality, and are therefore considered atomic relations of interdependence. When referring to cycles, we implicitly signify simple cycles, i.e., where all nodes are distinct[14]; we also discard self-loops and logical loops. When referring to conviviality, we refer to potential interaction, not actual interaction.*

H1 is based on two intuitions: (a) bridge rules represent potential ways of information exchange (actual information exchange occurs only when such rules are applied); and (b) self-loops, which are created by bridge rules that contain elements of the same context in their heads and bodies, and logical loops (e.g. a loop created by two bridge rules of the form: $r_1 = (C_1 : a) \leftarrow (C_2 : b)$ and $r_2 = (C_2 : b) \leftarrow (C_1 : a)$) do not actually enable information exchange between contexts, and should not therefore be taken into account when measuring conviviality.

H2 *Conviviality in a dependence network is evaluated in a bounded domain, i.e., over a [min, max] interval. This allows to read the values obtained by any evaluation method.*

This allows the comparison of different systems in terms of conviviality.

H3 *There is more conviviality in larger coalitions than in smaller ones.*

The intuition for H3 is that a greater number of collaborating contexts in a MCS offers a greater source of knowledge. This means that a large coalition of contexts can reach more informative conclusions or take more informative decisions compared to smaller coalitions.

H4 *The more coalitions in the dependence network, the higher the conviviality measure (ceteris paribus).*

H4 reflects the fact that the number of opportunities for information exchange for a context increases with the number of coalitions that the context participates in, which, in turn, increases with the number of bridge rules defined in this context.

Some coalitions provide more opportunities for their participating contexts to cooperate than others, being thereby more convivial. For a cooperative system modeled as MCS, the top goal should be to maximize conviviality. It should, therefore, fulfil the following two requirements:

R1 *Maximize the size of the coalitions, i.e., maximize the number of contexts involved in the coalitions.*

R2 *Maximize the number of these coalitions.*

Based on requirements $R1$ and $R2$, we define the *conviviality of a MCS M* as:

$$\Theta = \sum_{L=2}^{L=|C|} P(|C| - 2, L - 2) \times d_M^L, \tag{2}$$

$$\Omega = |C|(|C| - 1) \times \Theta, \tag{3}$$

$$\text{Conv}(M) = \frac{\sum_{c_i, c_j \in C, i \neq j} \text{coal}(c_i, c_j)}{\Omega} \tag{4}$$

where $|C|$ is the number of contexts in M, L is the cycle length, P is the usual permutation defined in combinatorics, $\text{coal}(c_i, c_j)$ for any distinct $c_i, c_j \in C$ is the number of cycles that contain the ordered pair (c_i, c_j) in $DN(M)$, such that the cycles do not represent logical loops, and Ω denotes the maximal number of pairs of contexts in cycles (which produces the normalization mentioned in Hypothesis H2). d_M is the maximum number of dependencies that a context in M may have on other contexts of M:

$$d_M = \max_{k \in M} \sum_{i=1}^{|C|} dep(k, c_i) \qquad (5)$$

Example 4.4. The dependence network of M, which is graphically represented in Figure 3 has three cycles: $\{(C_1, C_2, r_1), (C_2, C_1, r_3)\}$, $\{(C_1, C_3, r_2), (C_3, C_1, r_4)\}$ and $\{(C_1, C_3, r_2), (C_3, C_2, r_4), (C_2, C_1, r_3)\}$. The ordered pair (C_1, C_2) is only in the first cycle, therefore $coal(C_1, C_2) = 1$. In the same way we calculate $coal(C_2, C_1) = 2$, $coal(C_1, C_3) = 2$, $coal(C_3, C_1) = 1$, $coal(C_2, C_3) = 0$, $coal(C_3, C_2) = 1$. Following Equation 2 and assuming that $d_M = 1$, we calculate the conviviality of M as:

$$\mathrm{Conv}(M) = 7/\Omega = 0.58, \text{ where } \Omega = 12.$$

We note that $Conv(M)$ is almost maximal as adding only one bridge rule, namely from C_2 to C_3, results in a fully connected graph, i.e., maximal conviviality.

Computational complexity: For our measures, the number of cycles going through every possible pair of contexts is needed. The computational complexity for counting cycles can be computed using first the measures based on graph properties, that is in $O(|C|+|br_M|)$. Then, for each pair and cycle, a check must be performed to evaluate if the pair is in the cycle. Therefore, the complexity is $O((|C||C-1|)(|C|+|br_M|))$.

In the next section we show how one can use conviviality measures for MCS to compare different states of a distributed information system and improve it in terms of cooperativeness.

5 Inconsistency Resolution

As we previously argued, conviviality is a property that characterizes the cooperativeness of a MCS, namely the alternative ways in which the agents can share information in order to derive new knowledge. By evaluating conviviality, we are able to propose different ways in which cooperation can be increased, e.g., by suggesting new connections between the agents - or in other words mappings between their contexts. Consider, for example, a system in which an agent does not import data from any other agent. Recommending other agents from which the first agent can potentially import information from, can increase the conviviality of the system, which will in turn lead not only to enriching the local knowledge of the agent, but also the knowledge of the whole system.

5.1 Problem Description

Another way of using conviviality as a property of MCS, which we describe in more detail in this section, is for the problem of inconsistency resolution. In a MCS, even if contexts are locally consistent, their bridge rules may render the whole system inconsistent. This is formally described in [9] as a *lack of an equilibrium*. All techniques that have been proposed so far for inconsistency resolution are based on the same intuition: a subset of the bridge rules that cause inconsistency must be invalidated and another subset must be unconditionally applied, so that the entire system becomes consistent again. For nonmonotonic MCS, this has been formally defined in [15] as diagnosis:

"Given a MCS M, a *diagnosis* of M is a pair (D_1, D_2), $D_1, D_2 \subseteq br_M$, s.t. $M[br_M \backslash D_1 \cup heads(D_2)] \not\models \bot$". $D^{\pm}(M)$ is the set of all such diagnoses, while with $M[R]$ we denote the MCS obtained from M by replacing its bridge rules br_M with R; therefore $M[br_M \backslash D_1 \cup heads(D_2)]$ is the MCS obtained from M by removing the rules in D_1 and adding the heads of the rules in D_2.

In other words, if we deactivate the rules in D_1 and apply the rules in D_2 in unconditional form, M will become consistent. In a MCS it is possible that there is more than one diagnosis that can restore consistency.

Example 5.1. In our running example, consider the case that $prof B$ is also identified by C_2 as one of the authors of the paper under examination. In this case kb_2 would also contain $prof B$: $kb_2 = \{prof A, prof B\}$.

This addition would result in an inconsistency in kb_1, caused by the activation of rules r_4 and r_2. Specifically, rule r_4 would become applicable, *ubiquitousCom−puting* and *ambientComputing* would become *true* in C_3, r_2 would then become applicable too, and *distributedComputing* would become *true* in C_1 causing an inconsistency with *centralizedComputing*, which has also been evaluated as *true*. To resolve this conflict, one of the four bridge rules r_1-r_4 must be invalidated. Using the definition of diagnosis that we presented above, this is formally described as:

$$D^{\pm}(M) = \{(\{r_1\}, \emptyset), (\{r_2\}, \emptyset), (\{r_3\}, \emptyset), (\{r_4\}, \emptyset)\}.$$

Various criteria have been proposed for selecting a diagnosis including: *i.*) the number of bridge rules contained in the diagnosis - specifically in [15] pointwise subset-minimal diagnoses are preferred, *ii.*) local preferences on diagnoses proposed in [16] and *iii.*) local preferences on contexts and provenance information used in Contextual Defeasible Logic [2].

5.2 Proposed Solution

We propose using the conviviality of the resulted system as a criterion for selecting a diagnosis. This actually means that for each diagnosis, we measure the conviviality of the system that is derived after applying the diagnosis, and select the diagnosis that minimally decreases conviviality. The intuition is that the system should remain as *cooperative* as possible, and this is achieved by maximizing the amount of agents involved in the derivation of a conclusion or a decision and the number of potential ways in which a conclusion may be drawn. In the extreme case of invalidating all bridge rules, there will be no inconsistencies; however the agents will not able to take collective decisions - they will decide based on their local knowledge only. Overall, we propose resolving inconsistencies, by also keeping as many bridge rules (hence possibilities for information exchange) as possible.

Diagnoses contain two types of changes applicable in the bridge rules: invalidation (removal) of a rule; and applying a rule unconditionally, which means removing the body of the rule. These changes affect the dependencies of the system as follows: When invalidating or adding unconditionally rule r (as defined in (1)) in a MCS M, all the dependencies labeled by r are removed from the dependence network of M.

Assuming that $D_i = (D_{i1}, D_{i2})$ is a diagnosis that we can apply in a MCS M, and $M(D_i)$ is the MCS obtained M after applying D_i, the optimal diagnosis is the one that maximizes the conviviality of $M(D_i)$:

$$D_{opt} = \{D_i : \text{Conv}(M(D_i)) = max\}$$

Example 5.2. In the running example, there are four diagnoses that we can apply: D_1-D_4. Each of them requires invalidating one of rules r_1 to r_4, respectively. Figures 4-7 depict the four dependence networks $DN(M(D_i))$, which are derived after applying D_i. Dashed arrows represent the dependencies that are dropped in each $DN(M(D_i))$ compared to $DN(M)$.

Following Equation 2 and the four dependence networks (Figures 4-7) the conviviality of each DN is:

$$Conv(M(D_1)) = 5/\Omega = 0.42 \text{ and}$$
$$Conv(M(D_j)) = 2/\Omega = 0.17 \text{ with } j = 2, 3, 4 \text{ and } \Omega = 12$$

By applying D_1 (Figure 4), only one cycle $\{(C_1, C_2, r_1), (C_2, C_1, r_3)\}$ is removed from the initial dependence network $DN(M)$. However, by applying any of diagnoses D_2-D_4 (Figures 5-7), two cycles are removed from $DN(M)$. Therefore the optimal

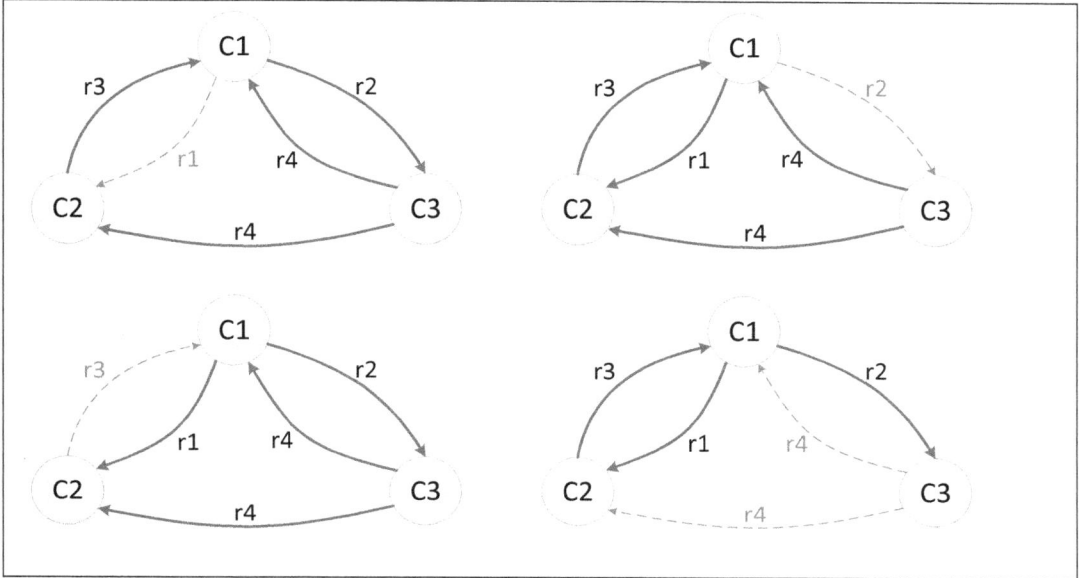

Figure 7: $DN(M(D_4))$

diagnosis is D_1. By applying D_1 the system will have the following equilibrium S':

$$S' = (\{sensors, corba, distributedComputing\},$$
$$\{profA, profB, middleware\},$$
$$\{ubiquitousComputing, ambientComputing\})$$

6   Related Research

The present work takes as a starting point the notion of social dependence and dependence graphs introduced by Castelfranchi and colleagues [13, 28], and further developed with a more abstract representation, similar to ours, in Boella et al. [4] and in the context of the concept of conviviality defined as reciprocity, in Caire et al. [10, 11]. Dependence based coalition formation is analyzed by Sichman [27], while other approaches are developed in [26, 17, 3].

Similarly to Grossi and Turrini [20], our approach brings together coalitional theory and dependence theory in the study of social cooperation within multiagent systems. However, our approach differs as it does not hinge on agreements, and that we extend it to MCS.

In section 5, we referred to three alternative approaches for resolving inconsistencies in MCS. Two of them are based on local preferences [16, 2]. Our approach differs in that we take into account a global property of the system, conviviality, with the goal of maximizing its cooperativeness. Our solution can be combined with any of these approaches. For example, one may choose to apply the conviviality-based approach only to those diagnoses that comply with some constraints representing user-defined criteria, as proposed in [16]. Another solution would be to define hybrid criteria, which combine preferences on diagnoses, either if these are explicitly defined as in [16] or if they are derived from preferences on contexts as in [2], with conviviality-based criteria. A study of such combined approaches will be part of our future work.

Our solution is more similar to the approach of [15], which selects the subset-minimal diagnosis: for pairs $A = \{A_1, A_2\}$, $B = \{B_1, B_2\}$, the pointwise subset relation $A \subseteq B$ holds iff $A_1 \subseteq B_1$ and $A_1 \subseteq B_2$. Conviviality-based resolution subsumes this approach, since, by definition, between two diagnoses D_1 and D_2, for which it holds that $D_1 \subseteq D_2$, it will always select D_1. Additionally, as we showed in section 5, it can also select between diagnoses that cannot be compared using this relation.

7 Conclusion

Today, with the rise of systems in which knowledge is distributed in a network of interconnected heterogeneous and evolving knowledge resources, such as the Semantic Web, Linked Open Data, and Ambient Intelligence, research in contextual knowledge representation and reasoning has become particularly relevant. Multi-Context Systems (MCS) are logical formalizations of distributed context theories connected through a set of bridge rules, which enable information flow between contexts. The individual agents, which are represented as contexts, cooperate by sharing information through their bridge rules. By combining and reasoning on the information they import, they are able to derive new knowledge. Evaluating the ways in which the system enables cooperations, and characterizing a MCS based on the opportunities for information exchange that it provides are therefore, key issues. The social science concept of conviviality has recently been proposed to model and measure the potential cooperation among agents in multiagent systems and ambient intelligence systems. Furthermore, formal conviviality measures for dependence networks using a coalitional game theoretic framework, have been introduced. Roughly, more opportunities to work with other agents increase the conviviality of the system.

This paper is a step towards extending the concept of conviviality to MCS. First, we describe how conviviality can be used to model cooperation in MCS. Based on the intuition that agents depend on the information they receive from other agents to achieve their goals (e.g. to take more informed decisions), we define dependence networks for MCS. Furthermore, the aim is for MCS to be as cooperative as possible, and for agents to have as many choices as possible to cooperate with other agents. This results in MCS being as convivial as possible. In order to evaluate conviviality, we apply pairwise conviviality measures and allow for comparisons among different MCS. Finally we propose a potential use of conviviality as a property of MCS for the problem of inconsistency resolution. In MCS, conflicts may arise as a result of importing mutually inconsistent knowledge from different contexts. Our approach is based on the idea that the optimal solution is the one that minimally decreases the conviviality of the system.

In further research, we contemplate the need to study alternative ways in which a MCS can be modeled as a dependence network. For example, another way to label dependencies among system contexts is to use the heads of the rules that these dependencies are derived from, instead of the rules themselves. This is based on the intuition that, the goal of applying a rule is actually to derive the conclusion that labels the head of the rule. This would require changing the definition of dependence networks to capture both disjunction (among rules that support the same conclusion) and conjunction (among the premises of each rule). We also plan to study the relation between the preference order on goals, which is included in the definition of dependence networks, and preferences on rules, contexts or diagnoses. Furthermore, we plan to combine the conviviality-based approach for inconsistency resolution with the preference-based approaches proposed by [16] and [2] and develop hybrid criteria for inconsistency resolution that take into account both local preferences and the conviviality of the system. Finally, we will study how the concept and tools for conviviality can be used in other distributed knowledge models, such as Linked Open Data, Distributed Description Logics [5], E-connections [22] and managed MCS [8], in which bridge rules are not only used to import information, but may also implement other operations, such as deletion or revision.

References

[1] Antoniou, G., Papatheodorou, C., Bikakis, A.: Reasoning about Context in Ambient Intelligence Environments: A Report from the Field. In: KR. pp. 557–559. AAAI Press (2010)

[2] Bikakis, A., Antoniou, G.: Defeasible Contextual Reasoning with Arguments in Ambient Intelligence. IEEE Trans. on Knowledge and Data Engineering 22(11), 1492–1506 (2010)

[3] Boella, G., Sauro, L., van der Torre, L.: Algorithms for finding coalitions exploiting a new reciprocity condition. Logic Journal of the IGPL 17(3), 273–297 (2009)

[4] Boella, G., Sauro, L., van der Torre, L.W.N.: Power and dependence relations in groups of agents. In: IAT. pp. 246–252. IEEE Computer Society (2004)

[5] Borgida, A., Serafini, L.: Distributed Description Logics: Assimilating Information from Peer Sources. Journal of Data Semantics 1, 153–184 (2003)

[6] Bouquet, P., Giunchiglia, F., van Harmelen, F., Serafini, L., Stuckenschmidt, H.: C-OWL: Contextualizing Ontologies. In: International Semantic Web Conference. pp. 164–179 (2003)

[7] Brewka, G., Eiter, T.: Equilibria in Heterogeneous Nonmonotonic Multi-Context Systems. In: AAAI. pp. 385–390 (2007)

[8] Brewka, G., Eiter, T., Fink, M., Weinzierl, A.: Managed Multi-Context Systems. In: IJCAI. pp. 786–791 (2011)

[9] Brewka, G., Roelofsen, F., Serafini, L.: Contextual Default Reasoning. In: IJCAI. pp. 268–273 (2007)

[10] Caire, P., Villata, S., Boella, G., van der Torre, L.: Conviviality masks in multiagent systems. In: 7th International Joint Conference on Autonomous Agents and Multiagent Systems (AAMAS 2008), Estoril, Portugal, May 12-16, 2008, Volume 3. pp. 1265–1268 (2008)

[11] Caire, P., Alcade, B., van der Torre, L., Sombattheera, C.: Conviviality measures. In: 10th International Joint Conference on Autonomous Agents and Multiagent Systems (AAMAS 2011), Taipei, Taiwan, May 2-6, 2011 (2011)

[12] Caire, P., Bikakis, A.: Enhancing Cooperation in Distributed Information Systems Using Conviviality and Multi-Context Systems. In: Multi-disciplinary Trends in Artificial Intelligence - 5th International Workshop, MIWAI 2011, Hyderabad, India, December 7-9, 2011. Proceedings. Lecture Notes in Computer Science, vol. 7080, pp. 14–25. Springer (2011)

[13] Castelfranchi, C.: The micro-macro constitution of power. Protosociology 18, 208–269 (2003)

[14] Cormen, T.H., Leiserson, C.E., Rivest, R.L., Stein, C.: Introduction to Algorithms. The MIT Press, 2nd edn. (2001)

[15] Eiter, T., Fink, M., Schüller, P., Weinzierl, A.: Finding Explanations of Inconsistency in Multi-Context Systems. In: Principles of Knowledge Representation and Reasoning: Proceedings of the Twelfth International Conference, KR 2010, Toronto, Ontario, Canada, May 9-13, 2010. AAAI Press (2010)

[16] Eiter, T., Fink, M., Weinzierl, A.: Preference-Based Inconsistency Assessment in Multi-Context Systems. In: Logics in Artificial Intelligence - 12th European Conference, JELIA 2010, Helsinki, Finland, September 13-15, 2010. Proceedings. Lecture Notes in Computer Science, vol. 6341, pp. 143–155. Springer (2010)

[17] Gerber, A., Klusch, M.: Forming dynamic coalitions of rational agents by use of the dcf-s scheme. In: AAMAS. pp. 994–995 (2003)

[18] Ghidini, C., Giunchiglia, F.: Local Models Semantics, or contextual reasoning=locality+compatibility. Artificial Intelligence 127(2), 221–259 (2001)

[19] Giunchiglia, F., Serafini, L.: Multilanguage hierarchical logics, or: how we can do without modal logics. Artificial Intelligence 65(1) (1994)

[20] Grossi, D., Turrini, P.: Dependence theory via game theory. In: van der Hoek, W., Kaminka, G.A., Lespérance, Y., Luck, M., Sen, S. (eds.) AAMAS. pp. 1147–1154. IFAAMAS (2010)

[21] Illich, I.: Tools for Conviviality. Marion Boyars Publishers, London (August 1974)

[22] Kutz, O., Lutz, C., Wolter, F., Zakharyaschev, M.: E-connections of abstract description systems. Artificial Intelligence 156(1), 1–73 (2004)

[23] Lenat, D.B., Guha, R.V.: Building Large Knowledge-Based Systems; Representation and Inference in the Cyc Project. Addison-Wesley Longman Publishing Co., Inc., Boston, MA, USA (1989)

[24] Parsons, S., Sierra, C., Jennings, N.R.: Agents that reason and negotiate by arguing. Journal of Logic and Computation 8(3), 261–292 (1998)

[25] Sabater, J., Sierra, C., Parsons, S., Jennings, N.R.: Engineering Executable Agents using Multi-context Systems. Journal of Logic and Computation 12(3), 413–442 (2002)

[26] Shehory, O., Kraus, S.: Methods for task allocation via agent coalition formation. Artif. Intell. 101(1-2), 165–200 (1998)

[27] Sichman, J.S.: Depint: Dependence-based coalition formation in an open multi-agent scenario. J. Artificial Societies and Social Simulation 1(2) (1998)

[28] Sichman, J.S., Conte, R.: Multi-agent dependence by dependence graphs. In: Procs. of The First Int. Joint Conference on Autonomous Agents & Multiagent Systems, AAMAS 2002. pp. 483–490. ACM (2002)

Received January 2013

FINITENESS OF PLAYS AND THE DIALOGICAL PROBLEM OF DECIDABILITY

NICOLAS CLERBOUT
University of Lille (UMR 8163 STL)
nicolas.clerbout@gmail.com

Abstract

We introduce the first study of a topic usually ignored when it comes to the metatheory of dialogical games, namely the dialogical problem of decidability. Our explanation and elucidation of the problem is done in dialogical terms only: it does not rest on equivalence results with other frameworks. Our analysis shows the decisive role in this matter of the mechanism of repetition ranks which was recently introduced to ensure finiteness of plays in dialogical games. The notion of repetition ranks thus turns out to be a fruitful and clarifying tool in metatheoretical studies on the dialogical framework.

Introduction

In dialogical[1] semantics, the meaning of expressions is given by the way they are used in an argumentative debate. The debate is designed as a game between two players called Proponent (**P**) and Opponent (**O**).

The dialogical approach has proven fruitful in relation to logic by giving a new perspective on the meaning of logical constants in terms of interaction within actual argumentative practices. Another aspect is that notions of interest in logic can adequately be captured in game-theoretical terms. The typical example is the notion of logical consequence who coincides with the existence of winning **P**-strategies in dialogical games. Winning strategies are a rather standard solution concept in mathematical games and the equivalence between this concept and logical consequence advocates for the conceptual benefits that can be expected from the game-based approach to semantics of the dialogical framework.

---

[1]The dialogical tradition is born in the late 1950s. It was introduced by P. Lorenzen who subsequently developed it with K. Lorenz. See Lorenzen&Lorenz [9].

In this paper we are interested in the topic of the decidability of this solution concept. That is to say : given a dialogical system[2] and an arbitrary formula φ, is there an effective way to correctly conclude in a finite amount of time whether there is a winning strategy for the Proponent in the game for φ?

To the best of our knowledge, this topic has not been addressed so far. On the one hand it is not really difficult to see why. Indeed when it comes to the dialogical approach to some logic, the existence of a winning **P**-strategy usually coincides with the notion of validity at stake. When there is such a coincidence result, dialogical decidability thus amounts to the decidability of the considered logical system. Since decidability is one of the basic properties investigated when studying a logical system, one may think that addressing the topic would be a lot of effort to rediscover well-known results. Indeed one can only expect, for example, the dialogical system for Propositional Logic to be decidable, or the one for First-Order Logic to be undecidable.

On the other hand however, this resulted in a lack of properly dialogical studies on such matters. In other words the study of this aspect of the metatheory of dialogical games has not really been introduced in a self-contained way. Instead, we have relied on coincidence results to export well-known results almost for free. Our motivation in this paper can thus be seen as a first step to remedy this situation. Indeed our aim is to consider the dialogical problem of decidability on its own: that is to say to formulate, understand and answer it in dialogical terms only. Such a task is not only interesting but also required. In the context of the project of developing the dialogical approach as an original theory of meaning of its own,[3] it is obviously of primary importance to study the properties of the framework. In particular the dialogical notion of (existence of) winning strategies is conceptually self-sufficient, and studies on its decidability should not be limited to the cases — no matter how numerous — where it coincides with notions from other approaches.

The main contribution of this paper is to identify clearly a feature of dialogical games inherently related to the problem of decidability, namely the mechanism of *repetition ranks*. The mechanism as introduced[4] and developed by Clerbout [1, 2] is rather simple: at the beginning of a play, the players each choose a positive integer — their ranks — which bound in a way to be explained below the number of moves

---

[2]By this we mean a set of rules which uniquely determines the dialogical game for a given thesis. See Section 1 for an example.

[3]See for example Rahman&Clerbout [11]

[4]Other presentations of the dialogical framework — such as Lorenz [7], Krabbe [6] or Rahman [10] — include a seemingly similar mechanism also called repetition ranks. Clerbout's account is different though and comes with the first direct and correct demonstration of the equivalence between tableaux and dialogues *observing the property of finiteness of plays*. For more on this topic, see Clerbout [2].

that will be available to them during the play.

Thus, repetition ranks ensure that every play ends after finitely many moves. By doing so, they ensure that the notion of victory is decidable.[5] In this paper we show that they are also strongly relevant when it comes to the question of the decidability of the notion of existence of winning **P**-strategies.

Indeed as we will explain hereafter, the difficulty is that there are in a single dialogical game infinitely many **P**-strategies. Moreover, in many cases the number of possible ways for **O** to play which a single **P**-strategy must take into account is also infinite. Hence the dialogical problem of decidability can be phrased in the following terms: is it or not possible to simplify the situation until we are able to accurately conclude in a finite amount of time that there is (or is not) a winning **P**-strategy in the game at hand? It is established in this paper that answering this question is always at least partly done by studying metatheoretical properties which relate the existence of winning **P**-strategy with repetition ranks.

The point is the following. Although their primary purpose is to ensure finiteness of plays, repetition ranks also introduce a form of infinity in dialogical games, namely in relation to the number of different plays in a dialogical game. That is because they are chosen by the players among the positive integers, which means that with ranks alone the players have infinitely many available ways to play. This is why it is necessary to consider them when we try to conclude that a dialogical system is decidable or not. Notice that in general it may not be enough to consider them since other features of a dialogical might introduce similar infinity in the number of plays. But there are also cases where the decidability or not of the system can be established only in terms of ranks-related properties of games. This is the case for example for dialogical games for Propositional and First-Order Logic which we will consider in this paper. Recall however that these unsurprising results are discussed as illustrations or applications of our analysis of the dialogical problem of decidability.

1 First-order dialogical games

We let $\mathcal{L}$ denote a first-order language (without equality) where every term is either a variable or an individual constant. A *move* is of the form $\mathbf{X}\,e$, where $\mathbf{X} \in \{\mathbf{O}, \mathbf{P}\}$ and e is of one of the following form:

-   Assertion: $!\varphi$, where φ is a sentence of $\mathcal{L}$.
-   Request: $?[A_0, \ldots, A_n]$, where each A_i is an assertion or a request.
-   $(\mathbf{n} := \mathbf{r}_i); (\mathbf{m} := \mathbf{r}_j)$, with $\mathbf{r}_i, \mathbf{r}_j \in \mathbb{N}^*$.

---

[5]See the structural rules and discussion at the end of Section 1.

The third possible form for e represents choices of repetition ranks among the positive integers (see the structural rules).

Local rules are triples of moves which show how assertions can be challenged and defended. They are given in Figure 1, where $\mathbf{X} \neq \mathbf{Y}$ and $\varphi(x/a_i)$ stands for the result of replacing every occurrence of x in φ by the individual constant a_i.[6]

Assertion	$\mathbf{X}\,!\varphi \vee \psi$	$\mathbf{X}\,!\varphi \wedge \psi$	$\mathbf{X}\,!\varphi \to \psi$	$\mathbf{X}\,!\neg\varphi$
Challenge	$\mathbf{Y}\,?[!\varphi,!\psi]$	$\mathbf{Y}\,?[!\varphi]$ or $\mathbf{Y}\,?[!\psi]$	$\mathbf{Y}\,!\varphi\,(?[!\psi])$	$\mathbf{Y}\,!\varphi$
Defence	$\mathbf{X}\,!\varphi$ or $\mathbf{X}\,!\psi$	$\mathbf{X}\,!\varphi$ resp. $\mathbf{X}\,!\psi$	$\mathbf{X}\,!\psi$	$--$

Assertion	$\mathbf{X}\,!\forall x\varphi$	$\mathbf{X}\,!\exists x\varphi$
Challenge	$\mathbf{Y}\,?[!\varphi(x/a_i)]$	$\mathbf{Y}\,?[!\varphi(x/a_1),\ldots,!\varphi(x/a_n),\ldots]$
Defence	$\mathbf{X}\,!\varphi(x/a_i)$	$\mathbf{X}\,!\varphi(x/a_i)$

Figure 1: Particle rules

A *play* is a sequence of moves which complies with the game rules. The *dialogical game* for a sentence φ is the set $\mathcal{D}(\varphi)$ of all plays with φ as the thesis (see SR0 below).

Structural rules define the conditions for a sequence of moves to be a play in a given dialogical game. The following notations are useful to formulate the rules. For every move M in a given sequence Σ of moves, $p_\Sigma(M)$ denotes the position of M in Σ. Positions are counted starting with 0. We also use below a function F, where the intended interpretation of $F_\Sigma(M) = [m', Z]$ is that in the sequence Σ, the move M is a challenge (if $Z = C$) or a defence (if $Z = D$) against the move of previous position m'.[7]

SR0 - Starting Rule. Let φ be a complex sentence of $\mathcal{L}$.[8] For any play $\Delta \in \mathcal{D}(\varphi)$ we have:

(i) $p_\Delta(\mathbf{P}\,!\varphi) = 0$,

(ii) $p_\Delta(\mathbf{O}\,n := \mathbf{r_1}) = 1$ and $p_\Delta(\mathbf{P}\,m := \mathbf{r_2}) = 2$ where $r_1, r_2 \in \mathbb{N}^*$.

SR1 - Classical Development Rule.

---

[6]We have added the request '$?[!\psi]$' between parentheses in the challenge against a material implication in order to make explicit the fact that $\mathbf{X}$ is expected to assert the consequent when defending. Otherwise it is arguably not clear why asserting the consequent should count as a defence at all. Aside from that and the notation, there is no difference with standard presentations of dialogical games such as Rahman&Keiff [12].

[7]This is inspired by and adapted from Felscher [3].

[8]The reason why atomic sentences are not included is related to SR2.

- For any move M in Δ such that $p_\Delta(M) > 2$ we have $F_\Delta(M) = [m', Z]$ where $Z \in \{C, D\}$ and $m' < p_\Delta(M)$.

- Let $\mathbf{r}$ be the repetition rank of Player $\mathbf{X}$ and $\Delta \in \mathcal{D}(\varphi)$ such that:
  - The last member of Δ is a $\mathbf{Y}$ move,
  - $M_0 \in \Delta$ is a $\mathbf{Y}$ move of position m_0,
  - There are n moves $M_1, \ldots, M_n$ of player $\mathbf{X}$ in Δ with $F_\Delta(M_1) = F_\Delta(M_2) = \ldots = F_\Delta(M_n) = [m_0, Z]$ and $Z \in \{C, D\}$.

  Let N be an $\mathbf{X}$ move such that $F_{\Delta^\frown N}(N) = [m_0, Z]$. We have $\Delta^\frown N \in \mathcal{D}(\varphi)$ if and only if $n < \mathbf{r}$.

SR2 - Formal Rule. The sequence Δ is a play only if the following condition is fulfilled: if $N = \mathbf{P} \,!\psi$ is a member of Δ, for any atomic sentence ψ, then there is a $M = \mathbf{O} \,!\psi$ in Δ such that $p_\Delta(M) < p_\Delta(N)$.

For our last structural rule we need the following definition:

Definition 1. Let Δ be a play in $\mathcal{D}(\varphi)$ the last member of which is an $\mathbf{X}$ move. If there is no $\mathbf{Y}$ move N such that $\Delta^\frown N \in \mathcal{D}(\varphi)$ then Δ is said to be $\mathbf{X}$ terminal.

SR3 - Winning Rule for Plays. Player $\mathbf{X}$ *wins* a play $\Delta \in \mathcal{D}(\varphi)$ if and only if Δ is $\mathbf{X}$ terminal.

Summing up, the structural rules set the following conditions. Any play in $\mathcal{D}(\varphi)$ starts with $\mathbf{P}$ asserting φ (the *thesis*). Then the two players choose their repetition ranks among the positive integers (SR0). After that, every move is a challenge or a defence of a previous move; players move alternately, and the number of challenges and defences they can perform in reaction to a same move is bounded by their repetition ranks (SR1).[9] $\mathbf{P}$ can assert an atomic sentence ψ only if $\mathbf{O}$ asserted it beforehand (SR2). The player who makes the last move of a terminal play wins it (SR3).

Let us present examples to illustrate the rules and give some additional explanations about the mechanism of repetition ranks. In the examples we consider atomic sentences φ and ψ.

---

[9]Hence any play is of finite length.

	O			P	
				$(\varphi \land (\varphi \to \psi)) \to \psi$	0
1	n := 1			m := 2	2
3	$\varphi \land (\varphi \to \psi)$ (?!ψ)	(0)		ψ	10
5	$\varphi \to \psi$		(3)	?!$(\varphi \to \psi)$	4
7	φ		(3)	?!φ	6
9	ψ		(5)	φ (?!ψ)	8

Explanations. Tables like this are a convenient way to represent plays because they make it easy to keep trace of the challenging and defending actions of the players. The outer columns give the positions of the moves within the play, and these are used to designate in the inner columns the moves which are challenged.

After the Proponent asserted the thesis, the players choose their repetition ranks with moves 1 and 2. Then the Opponent challenges the thesis by asserting the head and requiring the tail of the material implication. Since we have set ψ to be atomic, the Proponent cannot defend at once because of the formal rule $SR2$. He thus counter-attacks move 3 by requiring the two conjuncts with moves 4 and 6. Notice that because of $SR2$ again, **P** must play move 6 before he can challenge move 5 with his move 8. Once the Opponent asserts ψ as a defence in move 9, the Proponent can finally defend the thesis. Since there is no further possible move for **O** in this play, **P** wins it.

Notice that the Proponent can require both conjuncts with moves 4 and 6 only because he has chosen a repetition rank bigger than 1. Another important point about the mechanism of repetition ranks is that the structural rules we have given do not forbid what other presentations call *strict repetitions*. That is, the following sequence is perfectly admitted as a play by the rules we gave:

	O			P	
				$(\varphi \land (\varphi \to \psi)) \to \psi$	0
1	n := 1			m := 2	2
3	$\varphi \land (\varphi \to \psi)$ (?!ψ)	(0)			
5	$\varphi \to \psi$		(3)	?!$(\varphi \to \psi)$	4
7	$\varphi \to \psi$		(3)	?!$(\varphi \to \psi)$	6

Explanations. The play starts like the previous one, but this time **P** requires the same conjunct twice. Since he can challenge neither move 5 nor move 7 because of $SR2$, and since he has already challenged move 3 twice, the Proponent has no further possible move and **O** is the one who wins here.

These two plays illustrate the difference between winning a play and having a winning strategy. A smart Proponent will obviously play as in the first example and ensure victory. Coming back to the mechanism of repetition ranks, it is important to keep in mind that these positive integers are used to count the number of challenges and defences, regardless of the fact that the challenges or defences are the same. The reason why strict repetitions like move 6 in the second play are forbidden in other accounts of dialogical games is that they often are irrelevant in the sense that they cannot change the outcome of the play. But it is not the purpose of the rules to prevent the players from playing inefficiently.[10] The repetition ranks ensure finiteness of plays and that is all we require them to do.

We have mentioned in the Introduction that one of the effects of repetition ranks is that the notion of victory is decidable. This comes from the combination of rules $SR1$ and $SR3$. According to the latter victory is determined by looking at which player ends the play. Since every play is bound to end after finitely many moves because of repetition ranks, it is always possible to determine in a finite amount of time who wins a given play. Hereafter we investigate the relationship between ranks and the decidability of the notion of existence of winning **P**-strategies. We do so in a progressive way, and first introduce and explain the problem.

2   The leaf test

There is a quite obvious method which can *in principle* be used to discriminate between winning and non-winning strategies. The question of the decidability of a dialogical system then comes down to the extent to which this method can effectively be applied, and answers accurately in a finite amount of time to the question of the existence of winning **P**-strategies. We start by describing the method.

An **X**-*strategy* in a game $\mathcal{D}(\varphi)$ is a function which assigns a legal **X**-move to each non-terminal play the last move of which is a **Y**-move. An **X**-strategy is *winning* if playing according to it leads to **X**'s victory (in the sense of SR3) no matter how **Y** plays.

---

[10]Still, one might argue that it may be sensible to ban strict repetitions for the sake of avoiding some redundancy in the plays. However it is difficult to argue that such a strong view on redundancy can be uniformly adopted for any dialogical system we might be interested in.

Definition 2. By the *extensive form* $\mathfrak{E}(\varphi)$ *of the dialogical game* $\mathcal{D}(\varphi)$ we mean the tree representation of the game. The extensive form $\mathfrak{S}_x$ of an **X**-strategy s_x in $\mathcal{D}(\varphi)$ is the fragment of $\mathfrak{E}(\varphi)$ such that

(i)   the root of $\mathfrak{S}_x$ is the root of $\mathfrak{E}(\varphi)$.

(ii)   Suppose n is an **X**-node in $\mathfrak{S}_x$. Then any successor n' of n in $\mathfrak{E}(\varphi)$ is a successor of n in $\mathfrak{S}_x$.

(iii)   Suppose n is a **Y**-node in $\mathfrak{S}_x$. If n has at least a successor in $\mathfrak{E}(\varphi)$ then it has exactly one successor in $\mathfrak{S}_x$, namely the node associated with the **X**-move prescribed by s_x.

The following Lemma is then straightforward by virtue of SR3:

Lemma 1. *An* **X**-*strategy* s_x *in a given game* $\mathcal{D}(\varphi)$ *is winning if and only if every leaf in* $\mathfrak{S}_x$ *is labelled with an* **X**-*move.*

As a consequence, one can in principle easily determine whether a given **P**-strategy in a game $\mathcal{D}(\varphi)$ is winning. One simply needs to look at the leafs of the extensive form of the strategy and check whether they are all labelled with **P**-moves or not. We call this method *the leaf test* hereafter.

3   The obstacle of infinitary rules

Answering to the question of the decidability of a given dialogical system amounts to answering the following: to what extent can the leaf test be applied in the context of this system? Now in this respect we need to pay attention to the presence in the system of what we call infinitary rules. These are rules according to which a player has infinitely many available moves at some point during a play. The local rules for quantifiers, which we have recalled above, are of this kind. The point is that one of the player is to choose an individual constant for his challenge or defence — depending on the quantifier at stake. Moreover we usually consider a language with infinitely many individual constants, which makes these local rules infinitary.

A first consequence of the presence of infinitary rules is that we have to check infinitely many **P**-strategies with the leaf test. In our example of games for first-order languages, as soon as the Proponent gets to choose an individual constant he has in general infinitely many possible choices, hence infinitely many available strategies. This means that as long as we do not find a winning one, we have to keep using the leaf test, possibly forever. Moreover there is already a problem when we want to check even just one **P**-strategy which must take into account every possible way in which **O** might play. But when there are infinitary rules involved, there are an infinite number of such ways.

From these remarks it follows that in the presence of infinitary rules the dialogical problem of decidability amounts to the following question: when looking for the existence of a winning **P**-strategy, is there a conservative way in which the problem can be simplified? That is, is there a method by which we can safely — i.e., without loss of generality — reduce the task so that the problem is answered in a finite amount of time? More precisely: can we reduce the number and the "size" of **P**-strategies we need to check and accurately determine whether there is a winning one in the game at stake? A dialogical system is decidable exactly when the answer is yes.

It is now time to stress on the fact that, by the structural rule SR1, the players choose their repetition ranks among the positive integers. That is, this rule is infinitary and the very mechanism by which we ensure finiteness of plays raises the question we just formulated. This is the reason why, as we mentioned at the beginning of this study, the notion of repetition ranks is inherently related to the dialogical problem of decidability.

4   The reduction problem for ranks

When investigating the decidability of a dialogical system we must therefore always determine whether we can safely keep only a finite number of repetition ranks under consideration. We now present two cases where it is even enough to answer the reduction problem for ranks in order to know whether the system is decidable.

Indeed suppose SR1 is the only infinitary rule in the system at hand. Then the decidability or not amounts exactly to the reduction problem for ranks. An example is the dialogical approach to Propositional Logic. Given a standard propositional language, the rules of the associated dialogical system are the same as the one for Section 1 except that there is obviously no need for local rules for quantifiers.

We first notice that in propositional dialogical games it is enough to consider the case where the Opponent's repetition rank is 1:

> Let $\mathcal{D}^1(\varphi)$ denote the subgame of $\mathcal{D}(\varphi)$ where **O**'s rank is 1. There is a winning **P**-strategy in $\mathcal{D}(\varphi)$ if and only if there is one in $\mathcal{D}^1(\varphi)$.

The left-to-right direction is trivial. As for the interesting direction, we notice that while the Proponent is subjected to the formal rule SR2, there is no similar restriction for the Opponent. So let us assume that there is no winning **P**-strategy in $\mathcal{D}(\varphi)$. Then there is a winning **O**-strategy in the game.[11] This means that

---

[11]See Clerbout [1]: the Gale-Stewart Theorem [5] holds in the cases of dialogical games for

whenever the Opponent has the choice between several moves there is always at least one making it possible for her to win no matter what **P** does. Now since as noted above there is no particular restriction on **O**, she is always allowed to choose and play that winning move. Whether the move is a challenge or a defence, **O** simply needs rank 1 to be able to play it. Hence there is a winning **O**-strategy in $\mathcal{D}^1(\varphi)$. Therefore, if there is no winning **P**-strategy in $\mathcal{D}(\varphi)$ then there is not one in $\mathcal{D}^1(\varphi)$ either.

To our knowledge, this argument holds for almost every dialogical system which has been studied in the literature.[12] In particular, it also holds in the case of first-order dialogical games which we address below.

With this the leaf test can be applied to each **P**-strategy and determine in a finite amount of time whether it is winning or not since there is no other rule giving infinitely many choices to the Opponent. However, we still have infinitely many **P**-strategies to check — one for each positive integer as rank for the Proponent. This is the next point we need to address:

(Decidability of Propositional Dialogical Games)

*Let $\mathcal{D}^{1,2}(\varphi)$ denote the subgame of $\mathcal{D}^1(\varphi)$ where **P**'s rank is 2. There is a winning **P**-strategy in $\mathcal{D}^1(\varphi)$ if and only if there is one in $\mathcal{D}^{1,2}(\varphi)$.*

For the necessity part, it is enough to give an example for which **P** cannot win without a rank at least equal to 2. Consider any instance of $\neg(\varphi \wedge \neg\varphi)$ for an atomic φ:

	O				**P**	
					$\neg(\varphi \wedge \neg\varphi)$	0
1	n := 1				m := 2	2
3	$\varphi \wedge \neg\varphi$	(0)				
5	φ		(3)		? !φ	4
7	$\neg\varphi$		(3)		? !$\neg\varphi$	6
			(7)		φ	8

After move 8 there is nothing more the Opponent can do. But since φ is atomic, **P** needs **O** to assert it before in order to be allowed to play this winning move. This

---

Propositional and First-Order Logic because these are zero-sum well-founded games without tie. Hence in these games there is a winning strategy for exactly one of the players.

[12]The only exception we are aware of is the dialogical system for connexive logic, where the burden of the formal rule is sometimes transferred to the Opponent. See Rahman&Rückert [13].

is why in order to win he needs to require the first conjunct too i.e., to challenge move 3 twice. For that he needs his rank to be at least 2.

Now, apart from SR1, the rules of propositional dialogical games are at most binary: players have at most two different ways to challenge a given assertion, or to answer a given challenge. Obviously then, rank 2 is enough for **P** to be able to try all the different possibilities within one and the same play, and this establishes the other direction of the statement.

Summing up, in the case of propositional dialogical games, the number and size of **P**-strategies to check can be reduced. Once the Opponent's rank is set to be 1, each **P**-strategy has to take only finitely many different ways **O** can play into account. Moreover, we can restrict the search to a finite number of **P**-strategies, namely those which prescribe rank 2 for the Proponent. As a consequence, the leaf test can be used to accurately determine in a finite amount of time whether there is a winning **P**-strategy in a given game. Hence propositional dialogical games are decidable.

We now turn our attention to an other example and consider the reduction problem for ranks in the case of first-order dialogical games as defined in Section 1. As for the Opponent's rank, the same argument as previously holds and we are thus free to restrict our attention to $\mathcal{D}^1(\varphi)$. However, a quite simple example allows us to conclude the following about the Proponent's rank.

(Undecidability of First-Order Dialogical Games)

There is no positive integer **n** *such that, for any* φ, *there is a winning* **P***-strategy in* $\mathcal{D}^1(\varphi)$ *if and only if there is one in* $\mathcal{D}^{1,\mathbf{n}}(\varphi)$.

Thus, the reduction problem for ranks in first-order dialogical games gets a negative answer. This alone is enough to conclude that these games are undecidable. Indeed because of this as long as we do not find a winning **P**-strategy we have to try with a different rank for the Proponent and there are infinitely many to consider.

Let us present the example from which we can conclude the above statement about the Proponent's rank. Consider the formula *schema*

$$(\varphi_1 \wedge (\varphi_2 \wedge \varphi_3)) \rightarrow \psi$$

where :

φ_1 is $\forall x \forall y \forall z ((Rxy \wedge Ryz) \rightarrow Rxz)$
φ_2 is $\forall x \forall y ((Rxy \wedge Px) \rightarrow Py)$
φ_3 is $(Ra_1a_2 \wedge (Ra_2a_3 \wedge (\cdots \wedge (Ra_{k-1}a_k \wedge Pa_1)\ldots)))$
ψ is Pa_k

In the Appendix we consider the case where $k = 5$ and present a play from the dialogical game for this particular instantiation. The purpose is to provide an illustration which should make our next argument clearer and more precise.

The point is that, no matter the value of k, there is a winning **P**-strategy in the dialogical game considered. In order to ensure victory, the Proponent needs to use the transitivity of R (given by φ_1) *as many times as necessary* to get Ra_1a_k. After he has done so, he can use φ_2 together with Pa_1 (from φ_3) and thus get Pa_k to win. But the number of times necessary to get Ra_1a_k (i.e., the number of times **P** needs to challenge φ_1) obviously depends on k. In other words: the bigger k is, the bigger the rank **P** needs to win is.[13]

Let us write $\varphi[k]$ for the instance of the schema for a given k. Even if we eventually find the rank **n** such that there is a winning **P**-strategy in $\mathcal{D}^{1,\mathbf{n}}(\varphi[k])$, we will always have to try other, bigger ranks when looking for winning **P**-strategies in $\mathcal{D}(\varphi[k+1])$. So assume we take a rank for the Proponent as an upper limit beyond which we stop looking for winning **P**-strategies. Then for all the values of k for which the rank needed to win is bigger, we would wrongly conclude that there is no winning **P**-strategy.

We thus have found a collection of formulas for which there is no unique Proponent's rank to which we can reduce the question of the existence of winning **P**-strategies. This establishes the statement we made above and shows that first-order dialogical games are undecidable.

5 Concluding remarks: summary and semidecidability

The dialogical problem of decidability amounts to the possibility to actually apply the leaf test in order to get in a finite amount of time an accurate answer to the question of the existence of winning **P**-strategies. In this respect the task is to determine whether we can override infinitary rules and reduce to finite number and size the **P**-strategies to be tested.

While it ensures finiteness of plays, the mechanism of repetition ranks is implemented by means of the infinitary rule SR1. As such, it is therefore inherently connected to the decidability (or not) of a given dialogical system. Indeed when investigating the topic of decidability, one always needs to consider what we have called the reduction problem for ranks: this is part of establishing whether the leaf test can effectively be applied.

---

[13] In the example we give in Table 1, the Proponent needs to choose his rank $\mathbf{n} \geq 3$ in order to be able to win. The reader can easily check that there is no way to win with a smaller rank. Indeed with a smaller rank one of the moves 30, 38 or 46 would not be allowed to **P**.

In some cases the problem of decidability of a dialogical system actually boils down to the reduction problem for ranks. The most obvious case is when there is no infinitary rule except SR1 in the dialogical system at hand. A simple example is the dialogical approach to Propositional Logic where the reduction problem, and thus the question of the decidability of the system, gets a positive answer. The other case is when the reduction problem for ranks gets a negative answer: then we have at least one infinitary rule, namely SR1, which cannot be overlooked in a safe way. This immediately causes the system to be undecidable, and we saw an example with first-order dialogical games.

With these two examples we have given the dialogical manifestation of the decidability of Propositional Logic and of the undecidability of First-Order Logic. Moreover we have showed that these well-known properties are fully apprehended by considering the relationship between the existence of winning **P**-strategies and the mechanism of repetition ranks. That is, the concept of ranks does not only ensure finiteness of plays: it also brings some light on the dialogical problem of decidability and thus on the problem of decidability of a logic.

Let us come back one last time on first-order dialogical games. We have established in Section 4 that they are undecidable. But since the existence of winning **P**-strategies in this system adequately captures the notion of validity, one is entitled to expect a more fine-grained result of semidecidability matching the semidecidability of First-Order Logic. Let us therefore consider the formulas for which there is indeed a winning **P**-strategy. We show that this can be concluded in a finite amount of time: it is well-known that undecidability comes from some of the formulas which are not valid.[14] If there is a winning **P**-strategy in $\mathcal{D}(\varphi)$ then there is one in $\mathcal{D}^1(\varphi)$. From this it follows that there is an $\mathbf{n}$ such that there is a winning **P**-strategy in $\mathcal{D}^{1,\mathbf{n}}(\varphi)$: since there is a way for **P** to ensure victory, then there is at least one choice of rank for him which ensures victory after **O** chose rank $\mathbf{1}$.

The situation is thus: in those cases where there is a winning **P**-strategy in a dialogical game, we can in principle restrict our attention to a pair $(\mathbf{1}, \mathbf{n})$ of ranks for the players. But because the local rules for quantifiers are also infinitary rules, there are infinitely many **P**-strategies in $\mathcal{D}^{1,\mathbf{n}}(\varphi)$, each having to take infinitely many ways for **O** to play into account. However, there is a way to disregard most of the Proponent's strategies and most of the ways for the Opponent to challenge a universally quantified (or defend an existentially quantified) formulas. When a player is to choose an individual constant, he basically can choose between two kinds of constants: he can choose either a new one or one which already occurs in

---

[14]These are the formulas in what R. Smullyan [14] calls "the mystery class" of formulas. In model-theoretic terms, they are the formulas for which every counter-model has an infinite domain.

previous moves. Since finiteness of plays is ensured by the repetition ranks, there are necessarily only finitely many constants of the second kind. As for new constants we notice that even though there is an infinite number of these, we can say that they are in a way all equivalent modulo their name. By this we mean that the name of an individual constant has no impact when it comes to the possibility for a player to win. Thus, all the constants of the first kind are simply, so to say, alphabetical variants: the ramifications they open in the extensive form of the game are similar modulo the name of the constant. From this it follows that it is enough to consider one of them.

Summing up: we can always overlook most of the choices allowed to the players by the local rules for quantifiers. We only look at finitely many of these choices: one where the chosen constant is new, together with those where the constant occurs at some previous point of the play. By doing so, we have for each $\mathbf{n}$ only finitely many $\mathbf{P}$-strategies to check with the leaf test in $\mathcal{D}^{1,\mathbf{n}}(\varphi)$, each accounting for only finitely many ways for $\mathbf{O}$ to play. It follows that if there is an $\mathbf{n}$ such that there is a winning $\mathbf{P}$-strategy in $\mathcal{D}^1(\varphi)$, we will eventually be able to establish it after checking the $\mathbf{n}-1$ previous repetition ranks for the Proponent. That is to say, first-order dialogical games are semidecidable. Of course if there is no such $\mathbf{n}$, then this procedure will never stop and that is why these games are undecidable.

References

[1] Clerbout, N.: 2013a, 'First-Order Dialogical Games and Tableaux'. *Journal of Philosophical Logic*. Online First Publication. DOI: 10.1007/s10992-013-9289-z.

[2] Clerbout, N.: 2013b, *Etude sur quelques sémantiques dialogiques. Concepts fondamentaux et éléments de metathéorie*. PhD Dissertation, Universities of Leiden and Lille.

[3] Felscher, W.: 1985a, 'Dialogues, Strategies, and Intuitionnistic Provability'. *Annals of Pure and Applied Logic* **28**, 217–254.

[4] Felscher, W.: 1985b, 'Dialogues as a Foundation for Intuitionnistic Logic'. In: D. Gabbay and F. Guenthner (eds.): *Handbook of Philosophical Logic. Volume 3: Alternatives in Classical Logic*, Vol. 166 of *Studies in Epistemology, Logic, Methodology, and Philosophy of Science*. Dordrecht/Hingham: Kluwer, pp. 341–372.

[5] Gale, D. and Stewart, F. M.: 1953, 'Infinite games with perfect information'. In: H. W. Kuhn and A. W. Tucker (eds.): *Contributions to the Theory of Games, volume II*, Vol. 28 of *Annals of Mathematics Studies*. Princeton: Princeton University Press, pp. 245–266.

[6] Krabbe, E. C.: 1985, 'Formal Systems of Dialogue Rules'. *Synthese* **63**, 295–328.

[7] Lorenz, K.: 1968, 'Dialogspiele als semantische Grundlage von Logikkalkülen'. *Archiv für mathematische Logik und Grundlagenforschung* **11**, 32–55 and 73–100. Reprinted in [9].

[8] Lorenz, K.: 2001, 'Basic Objectives of Dialogue Logic in Historical Perspective'. *Synthese* **127**, 255–263.

[9] Lorenzen, P. and Lorenz, K.: 1978, *Dialogische Logik*. Darmstadt: Wissenschaftliche Buchgesellschaft.

[10] Rahman, S.: 2012, 'Negation in the Logic of First Degree Entailment and *Tonk*. A Dialogical Study'. In: S. Rahman, M. Marion and G. Primiero (eds.): *The Realism-Antirealism Debate in the Age of Alternative Logics*. Dordrecht: Springer, pp. 157–201.

[11] Rahman, S. and Clerbout, N.: 'Constructive Type Theory and the Dialogical Turn. A New Start for the Erlanger Konstruktivismus'. forthcoming

[12] Rahman, S. and Keiff, L.: 2005, 'On How to Be a Dialogician'. In: D. Vanderveken (ed.): *Logic, Thought and Action*. New-York: Springer, pp. 359–408.

[13] Rahman, S. and Rückert, H.: 2001, 'Dialogical Connexive Logic'. *Synthese* 127(1-2), pp. 105–139.

[14] Smullyan, R.: 1968, *First-Order Logic*. New York: Springer Verlag.

Appendix

In this appendix we consider an instance of the schema discussed in Section 4 where $k = 5$. To be more specific we consider in Table 1 a play from the dialogical game. For the sake of readability, we have used the same abbreviations φ_1, φ_2 and φ_3 when possible, and in challenges against material implication we have indicated only the assertion of the head.

Although quite long, this play is ultimately rather simple and it should be clear from it that there is a winning **P**-strategy in the game. Also, we hope it is helpful to understand and illustrate the reasoning which led us to conclude to the undecidability of first-order dialogical games.

	O			P	
				$[\forall x\forall y\forall z((Rxy \wedge Ryz) \to Rxz)$ $\wedge(\forall x\forall y((Rxy \wedge Px) \to Py)$ $\wedge(Rab \wedge (Rbc \wedge (Rcd \wedge (Rde \wedge Pa)))))]$ $\to Pe$	0
1	n := 1			m := 3	2
3	$[\forall x\forall y\forall z((Rxy \wedge Ryz) \to Rxz)$ $\wedge(\forall x\forall y((Rxy \wedge Px) \to Py)$ $\wedge(Rab \wedge (Rbc \wedge (Rcd \wedge (Rde \wedge Pa)))))]$	(0)		Pe	62
5	$\forall x\forall y\forall z((Rxy \wedge Ryz) \to Rxz)$		(3)	$?[!\varphi_1]$	4
7	$(\forall x\forall y((Rxy \wedge Px) \to Py)$ $\wedge(Rab \wedge (Rbc \wedge (Rcd \wedge (Rde \wedge Pa)))))$		(3)	$?[!\varphi_2 \wedge \varphi_3]$	6
9	$\forall x\forall y((Rxy \wedge Px) \to Py)$		(7)	$?[!\varphi_2]$	8
11	$Rab \wedge (Rbc \wedge (Rcd \wedge (Rde \wedge Pa)))$		(7)	$?[!\varphi_3]$	10
13	Rab		(11)	$?[!Rab]$	12
15	$Rbc \wedge (Rcd \wedge (Rde \wedge Pa))$		(11)	$?[!Rbc \wedge (Rcd \wedge (Rde \wedge Pa))]$	14
17	Rbc		(15)	$?[!Rbc]$	16
19	$Rcd \wedge (Rde \wedge Pa)$		(15)	$?[!Rcd \wedge (Rde \wedge Pa)]$	18
21	Rcd		(19)	$?[!Rcd]$	20
23	$Rde \wedge Pa$		(19)	$?[!Rde \wedge Pa]$	22
25	Rde		(23)	$?[!Rde]$	24
27	Pa		(23)	$?[!Pa]$	26
29	$\forall y\forall z((Ray \wedge Ryz) \to Raz)$		(5)	$?[!\varphi_1(x/a)]$	28
31	$\forall z((Rab \wedge Rbz) \to Raz)$		(29)	$?[!\forall z((Rab \wedge Rbz) \to Raz)]$	30
33	$(Rab \wedge Rbc) \to Rac$		(31)	$?[!(Rab \wedge Rbc) \to Rac]$	32
37	Rac		(33)	$Rab \wedge Rbc$	34
35	$?[!Rab]$	(34)		Rab	36
39	$\forall z((Rac \wedge Rcz) \to Raz)$		(29)	$?[!\forall z((Rac \wedge Rcz) \to Raz)]$	38
41	$(Rac \wedge Rcd) \to Rad$		(39)	$?[!(Rac \wedge Rcd) \to Rad)]$	40
45	Rad		(41)	$Rac \wedge Rcd$	42
43	$?[!Rac]$	(42)		Rac	44
47	$\forall z((Rad \wedge Rdz) \to Raz)$		(29)	$?[!\forall z((Rad \wedge Rdz) \to Raz)]$	46
49	$(Rad \wedge Rde) \to Rae$		(47)	$?[!(Rad \wedge Rde) \to Rae]$	48
53	Rae		(49)	$Rad \wedge Rde$	50
51	$?[!Rad]$	(50)		Rad	52
55	$\forall y((Ray \wedge Pa) \to Py)$		(9)	$?[!\varphi_2(x/a)]$	54
57	$(Rae \wedge Pa) \to Pe$		(55)	$?[!(Rae \wedge Pa) \to Pe]$	56
61	Pe		(57)	$Rae \wedge Pa$	58
59	$?[!Rae]$	(58)		Rae	60

Table 1: **P** needs rank ≥ 3 to win

Received January 2014

Systems of Interacting Argumentation Networks

Dov Gabbay
Bar-Ilan University, Israel
King's College London, UK
University of Luxembourg, Luxembourg
University of Manchester, UK
dov.gabbay@kcl.ac.uk

Abstract

This paper offers semantical options for interacting families of abstract argumentation networks. Such families arise in a variety of contexts and application areas:

1. Several agents come each with their own network and complete Dung extensions and they are trying to merge them into a consensus common network with its own complete extensions

2. Fibred network of networks, where for reasons arising from an application area we substitute an entire network for a specific argument node in another network. This is a form of instantiation, not with proofs (as in ASPIC), but with entire networks.

3. Some semantics, such as CF2, regards a network as an acyclic directional network of maximal loops (maximal strongly connected components) and defines the semantics on such a system.

4. Argumentation interaction between several conflicting (i.e. attacking) world points of view, such as liberal socialist point of view vs. religious fundamentalism. In such cases merging of their respective networks is not practical nor meaningful, and other semantical measures are required.

5. A recursive combination of all the above.

We offer a general framework for handling such complex systems and discuss at length our possible semantic options.

Research supported by the Israel Science Foundation Project 1321/10: Integrating Logic and Networks.

1   Introduction

In this section we motivate our semantics and explain its ideas. Consider the following true story. I was, at the time, Head of Department of Mathematics and Computer Science. Part of my job was to identify members of staff ready for academic promotion, collect references on the candidates, prepare a file on those candidates who have a strong case for promotion and pass the file on to the Dean with a positive recommendation. The Dean then decided whether the case merited support and passed the file on to the Rector for actually executing the promotion.

The story begins when three candidates were put forward by me to the Dean to be promoted to Professorships. The Dean asked me for a meeting and said that he agreed that these three candidates deserved promotion on objective international academic considerations, but alas, he had budgetary problems and cuts and could not afford the extra salary payments required by promotions across all departments in the faculty. Therefore, he added, that he had decided that there would be no promotions round at all that year.

I got very upset and said that money considerations should not interfere with academic considerations and that he should promote the candidates but, if necessary, not raise their salaries. The Dean said that he could not do that.

I put the cases forward again the following year, and the Dean again said that he could not promote the candidates for financial reasons. In fact, he said, the financial situation was then even more difficult than the year before. This time I was more prepared for this argument and said that deciding on the faculty budgets and cuts is a matter of balance and priorities and I challenged the Dean by saying "move over, let me run the faculty in your place". The Dean was taken aback and said that he would see what he could do. As far as I remember, he promoted one of the candidates.

The argumentation analysis of this story is in terms of Bench-Capon's value based argumentation networks [20], as seen in Figure 1.

In the first meeting I claimed that we should not accept attacks from frame 1 to frame 2.

This claim did not help. In the second round, I asked the Dean to move over and let me build an extension (i.e. budgetary allocation in frame 1) that would not attack the promotion in frame 2. This was the only way to defend the promotion. In the light of my threat, the Dean revised his extension for frame 1 and managed to make one promotion.

Another example of this type is when a theological system attacks argument x, put forward by a secular social system. It is not practical to say "I do not accept your religion; it is not relevant to attack secular considerations by a fabricated

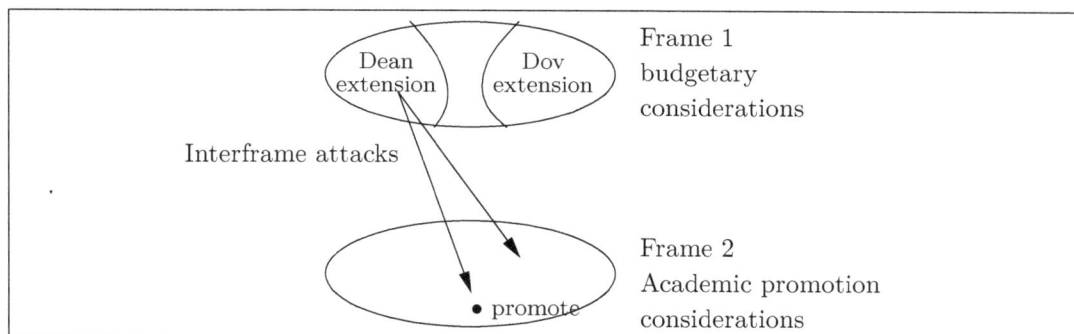

Figure 1:

belief system". It is much more effective to get into the theological considerations themselves and argue that there is a theological point of view (an extension) which does not attack x. Equally and symmetrically, it is no use for the religious system to dismiss the social considerations as irrelevant because they go against the will of God. It is much better to go into the social arguments and adopt a religiously more tolerable extension. In fact both the religious argumentation system and the secular social system may be the result of merging several second level subsystems of relgous factions and political parties respectively. In such a case one side needs to address the internal factions of the other side with the hope of favourably influencing the final merged outcome. Section 3 will deal with such higher level situations.

Politicians know this. If they are criticised by the Church or by some Nobel prize winner economists, or by some experts, it is best to get another expert or another Nobel prize winner or another bishop to support them!

Another example is from the legal domain. Suppose we build a prosecution case against a defendant. Call this Frame 2. This case contains a piece of evidence p, obtained by some questionable procedures. Frame 1 is the considerations from the legal theory of evidence to check whether it is possible to admit p. The prosecution will adopt one extension and the defence will adopt a different extension. See [23].

Let us now see what kind of formal semantics this way of thinking requires. Let us look at Figure 2.

This is a directional value based system. There are four frames (these are the "values"). The frames names are $V = \{v_1, \ldots, v_4\}$ and the binary ordering on V is as in Figure 2. The frames attack each other in a directional way (no cycles, i.e.' we do not have x, y in frame i and u, v in frame j, with $i \neq j$ and x attacks u and v attacks y).[1]

―――――――――――――――――――――

[1]This is a harsh restriction. We shall look at loops in Section 3. We need to present the basic model first in this section.

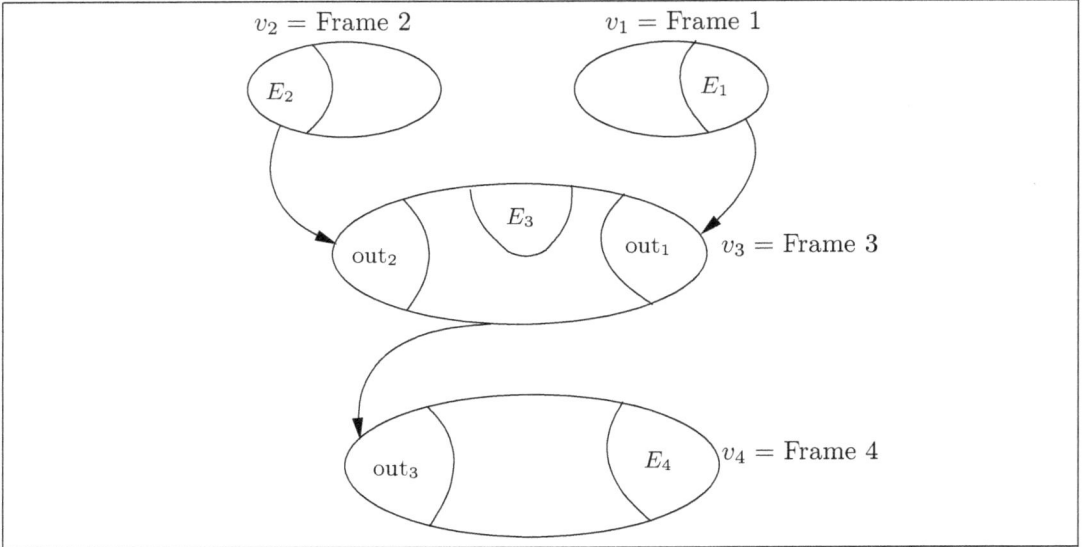

Figure 2:

The semantical considerations may differ from one frame to another. So for example frame 1 may take grounded extensions while frame 2 may take preferred extensions. We need not take proper Dung extensions at all. Any selection of conflict free sets can do. So maybe we should call our choices "coherent views".[2] Let us put forward extensions ("coherent views") E_1 and E_2 in frame 1 and frame 2 respectively. These extensions attack elements in frame 3 and "take out" the nodes in out_1 and out_2 respectively. What remains of frame 3 (i.e. frame 3 $-$ $\mathrm{out}_1 \cup \mathrm{out}_2$)) can now have extension E_3. E_3 attacks frame 4 and leaves it with frame 4 $-$ out_3. and frame 4 semantics would yield the extension E_4 of frame 4 $-$ out_3.

So the total extension of $\bigcup_i$ Frame i is $\bigcup_i E_i$ and this is calculated as described.

Note that the set of frames V can be considered itself as an argumentation network with a ground extension $\{v_1, v_2, v_4\}$, but we do not look at V in this way and do not take extensions. We do not use this option. Thus Frame 3 is not ignored and considered "out".

Example 1.1. *Figure 3 is an example:*

> *Frame 1* $= (\{x\}, \{(x, x)\})$.
> *Frame 2* $= (\{a, b, c\}, \{(a, b), (a, c), (c, a)\})$.

---

[2]Indeed this is going to be the case in Section 3.3, when we give a general Definition of fibred argumentation systems and define extensions for them in Definition 3.12.

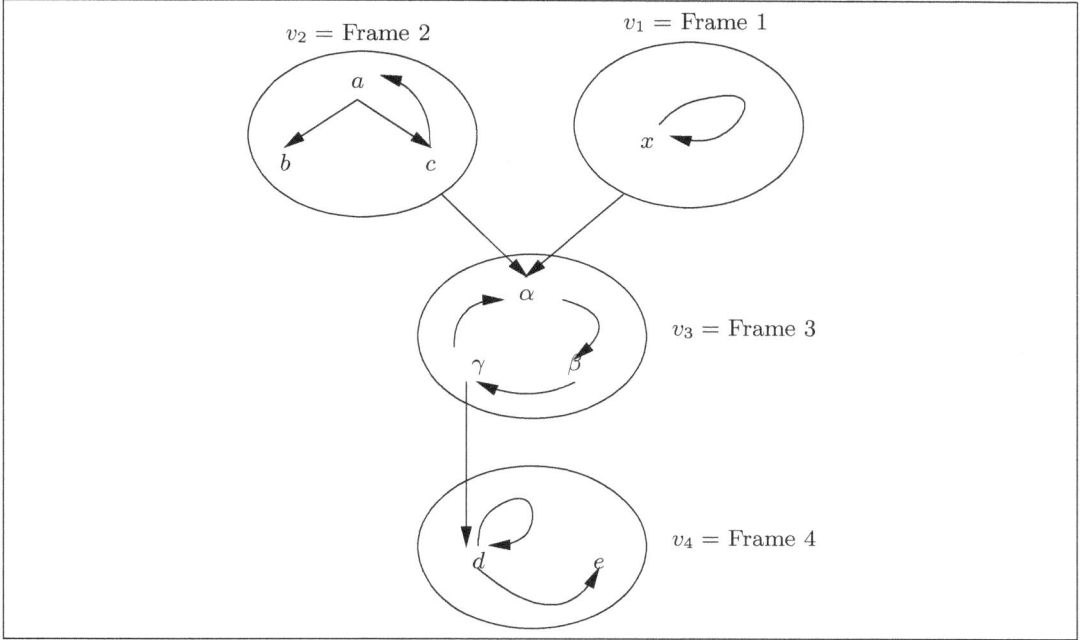

Figure 3:

Frame 3 = $(\{\alpha, \beta, \gamma\}, \{\alpha, \beta), (\beta, \gamma), (\gamma, \alpha)\}$
Frame 4 = $(\{d, e\}, \{(d, d), (d, e)\}$.
Interframe attacks are $\{(c, \alpha), (x, \alpha), (\gamma, d)\}$.

Let us take the semantics of complete extensions in any frame. We now show one possible directional extension.

Step 1: Choose extensions in the top frames; frame 1 and frame 2.

> Frame 1: Take $E_1 = \varnothing$
> Frame 2: Take $E_2 = \{c, b\}$.

Step 2: We now replace the frames 1 and 2 by the extensions. We get Figure 4.

Step 3: We propagate the attacks and look at the remainder graph of remainder frames. We get Figure 5. Note that the node α was taken out by the attack from node c.

Step 4: (This is a "goto step 1" recursion operation on Figure 5.)

> We now take an extension $E_3 = \{\beta\}$ in frame 3. We get Figure 6.

135

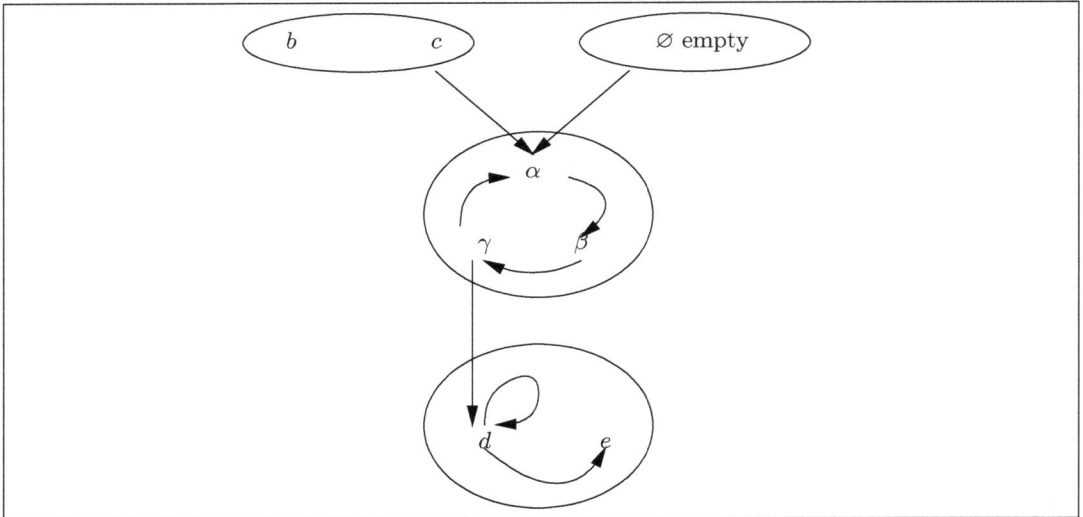

Figure 4:

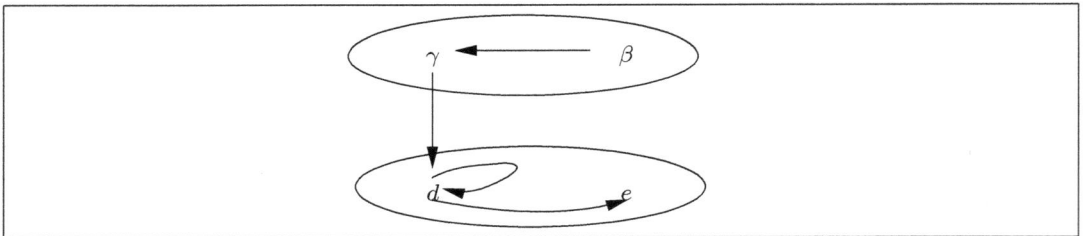

Figure 5:

Step 5: *We now take extension in frame 4, $E_4 = \varnothing$.*

The total directional extension of Figure 3 is $\{c, b, \beta\}$.

Had we chosen the extension $\{a\}$ in frame 2 we would have got $\{a\}$ as a total extension.

So the family of all extensions for Figure 3 is $\{c, b, \beta\}$ and $\{a\}$.

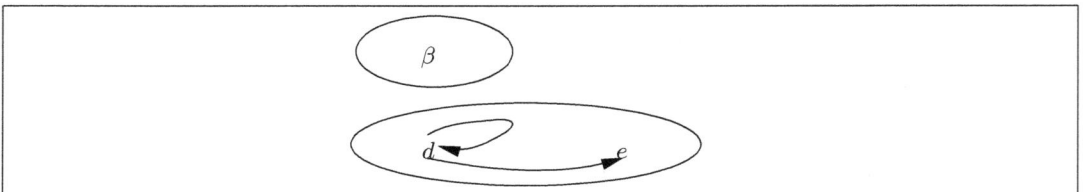

Figure 6:

136

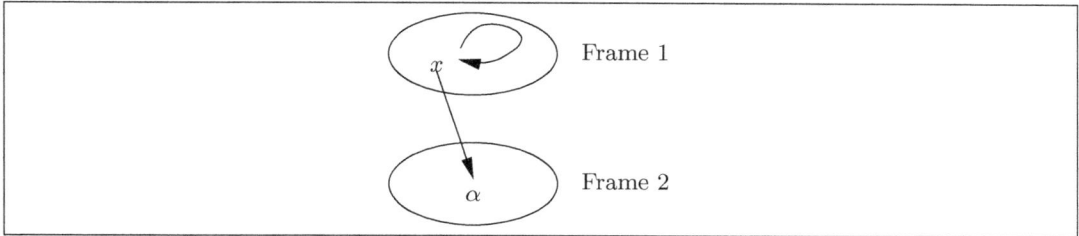

Figure 7:

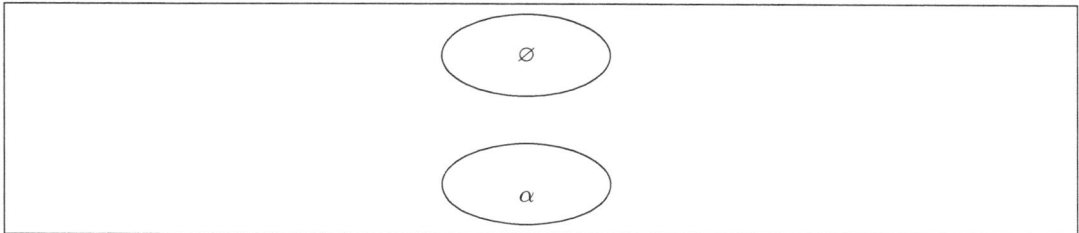

Figure 8:

Example 1.2. *Consider Figure 7.*

The extension of frame 1 is $\varnothing$. So we get Figure 8.

The total directional extension is $\{\alpha\}$. If we look at Figure 7 as a single frame we get $x = \alpha =$ undecided and the extension is $\varnothing$.

The reader can see a connection with the CF2 semantics, but the approach and philosophy involved are different.

Remark 1.3. *Note that the directional extensions depend on our choice of values and for each value the choice of "coherent views" (extensions). Consider Figures 9 and 10.*

Let us adopt the policy of taking at each frame as "coherent views" (or "coherent positions") maximal conflict free sets..

Note that this is not the CF2 semantics, because in the directional semantics we

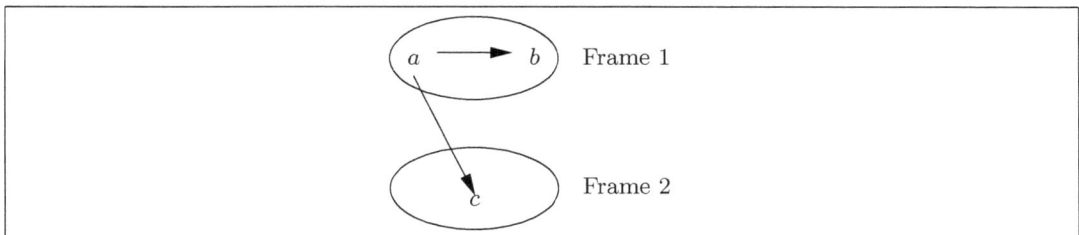

Figure 9:

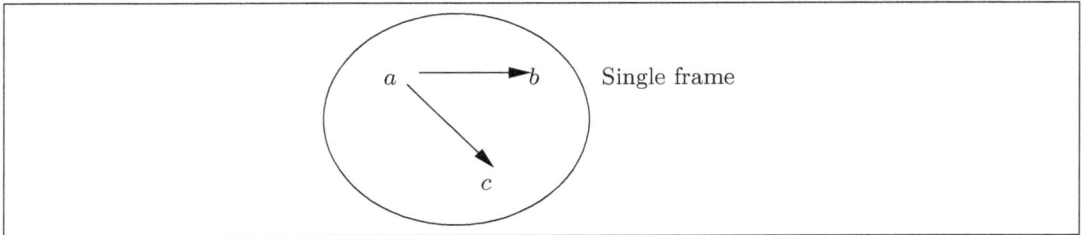

Figure 10:

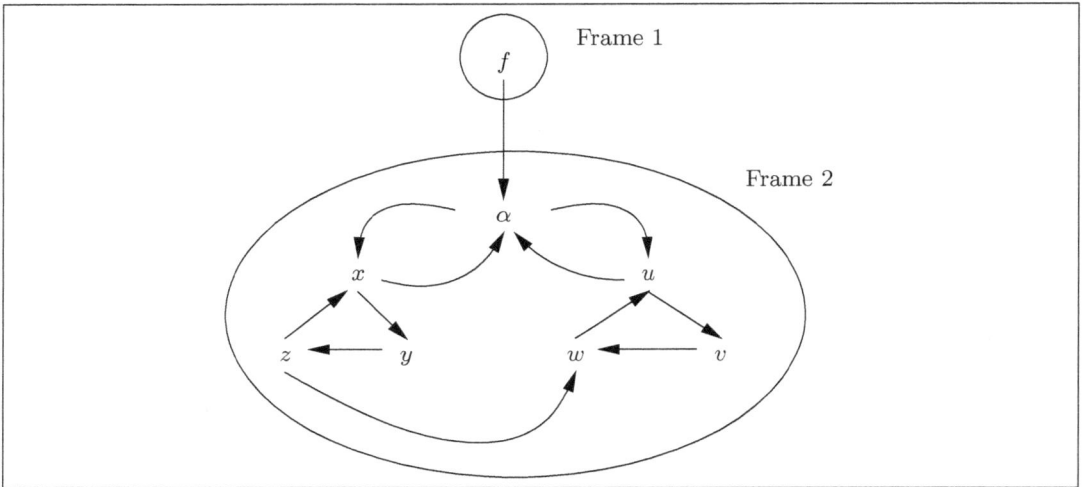

Figure 11:

take maximal conflict free subsets in all cases, even the cases where there are no loops. Call this semantics CFD semantics.

So in Figure 10 we get two possible CFD "extensions", $\{a\}$ and $\{b,c\}$. We get $\{b,c\}$ even though there are no loops. CF2 would get only $\{a\}$.

In Figure 9, we need to go through two steps. First calculate all CFD extensions for the first frame 1. This yields $\{b\}$ and $\{a\}$. For $\{a\}$ we get $\varnothing$ in frame 2. For $\{b\}$ we get $\{b,c\}$ as an extension. So the total CFD extensions are $\{a\}$ and $\{b,c\}$. This is the same as for Figure 10 but this is a coincidence.

Example 1.4. *We would like to have the CF2 semantics as a special case of our CFD semantics, where we take maximal conflict free subsets as our coherent "extensions". Can we do that or do we need a slight change in concepts?*

Consider Figure 11.

Frame 1 and frame 2 are also the maximal strongly connected subsets of the total set of nodes and attacks. Both the CF2 semantics and the CTD semantics will use

138

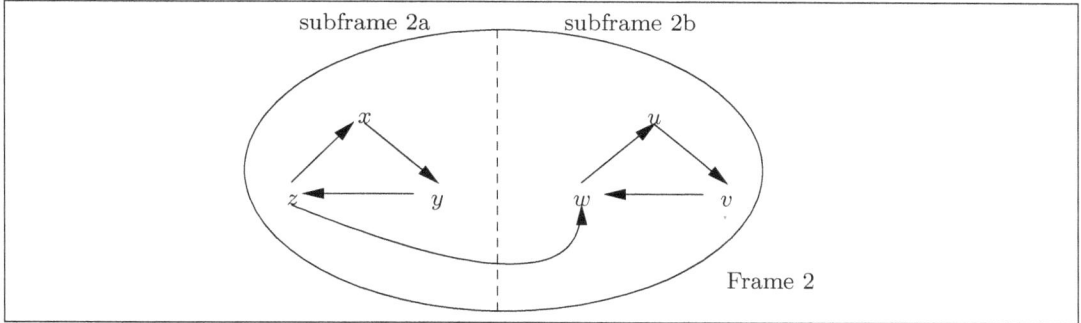

Figure 12:

f to attack frame 2 and end up with Figure 12.

To get the CF2 extension, we need to divide frame 2 into subframe 2a and subframe 2b (see [21], where there is a representation theorem for computing CF2 semantics in a directional way, as we are doing here).

The CFD semantics will not divide frame 2 but take maximal conflict free subsets of frame 2.

Thus the CFD semantics will take $\{z, v\}$ as an extension in frame 2, thus allowing for the final "extension" $\{f, z, v\}$. The CF2 semantics, having chosen $\{z\}$ first as a maximal conflict free subframe 2a, will be left with $\{u\}$ only as an extension of subframe 2b. Thus $\{f, z, v\}$ is not a CF2 extension. So what is the methodological difference between CFD and CF2?

The difference, according to what is proved in [21] and using the notation and approach of this paper, is that CF2 continues to value annotate the remainder frame (in our case frame 2 of Figure 12) and divide it into subframes. We shall thus adopt this principle in our general definitions so that our CFD semantics can contain the CF2 semantics as a special case.

The reader should compare with the abstract work of Ringo Baumann on splitting [28, 29]. There are points of contact but only formally.

2 Directional context semantics

We now give formal definitions for our concepts and algorithms.

Definition 2.1. *A general (resp. directional) value based argumentation framework has the form*

$$
\begin{aligned}
\mathcal{A} &= (S, R, V, <, \mathbf{e}) \\
&= (S, R, \mathbb{V}, \mathbf{e}) \\
with\ \mathbb{V} &= (V, <)
\end{aligned}
$$

where S is a set of arguments, $R \subseteq S \times S$ is the attack relation and $\mathbb{V} = (V, <)$ is a general binary (resp. acyclic) graph of values and $\mathbf{e}$ is a function from S into V such that the following condition holds:

(*) whenever we have xRy then $\mathbf{e}(y) \leqq \mathbf{e}(x)$.[3]

Figures 13 and 14 are examples. Figure 13 is directional while Figure 14 is general value based, but is not directional.

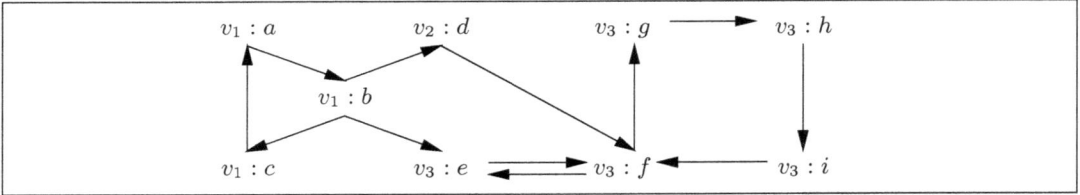

Figure 13:

Definition 2.2. Let $\mathcal{A}$ be a directional value based network. Let $v \in V$. Define as $\mathcal{A}_v = (S_v, R_v)$, where
$$S_v = \{s \in S, \mathbf{e}(s) = v\}$$
$$R_v = R \upharpoonright S_v$$

Definition 2.3. A universal value function is a function or an algorithm $\mathbb{A}$ such that given a finite argumentation frame (S, R) would yield a directional value based network $\mathcal{A}$ based on (S, R) of the form $\mathcal{A} = (S, R, V, <, \mathbf{e})$ (in other words, $\mathbb{A}$ would supplement (S, R) with $(V, <, \mathbf{e})$).

We require that $\mathbb{A}$ satisfies the following refinement condition:

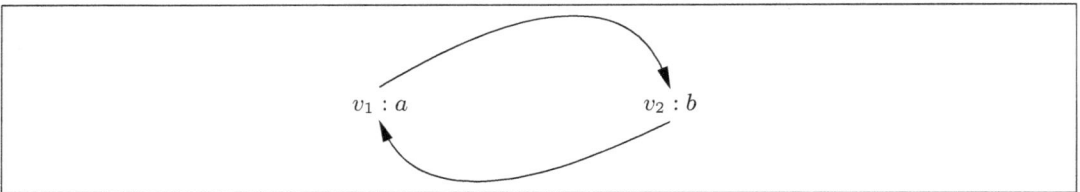

Figure 14:

---

[3]This condition (*) forces our interpretation on the family of argumentation networks to be directional. Take the extreme case where the ordering on V is empty. Then without this condition we have a case of merging networks, with V interpreted as the set of voters and with $<$ being some importance priority among the voters. The received wisdom in the argumentation community for such a merging example is to use some form of voting to obtain an extension of the system. We compare and discuss these fine points in Section 3.3 below.

140

If (S', R') is a subnetwork of (S, R), (i.e. S' is a subset of S and R' is a subset of R) and $(V', <', \mathbf{e}')$ and $(V, <, \mathbf{e})$ are the respective valuation systems then $(V', <', \mathbf{e}')$ is a refinement of $(V, <, \mathbf{e})$, namely for x, y in S if $\mathbf{e}'(x) = \mathbf{e}'(y)$ then $\mathbf{e}(x) = \mathbf{e}(y)$.

Example 2.4. *Consider the following universal value function $\mathbb{C}$ defined as follows. Given an argumentation network (S, R), let $(V, <)$ be the acyclic network (obtained from it) of the maximal strongly connected subsets of Baroni [1]. Let $\mathbf{e}(x) = x^* \in V$, where x^* is the maximal strongly connected set containing x (or is $\{x\}$ itself if x is not part of any loop). Define $x^* < y^*$ iff yRx. We have $x^* < y^*$ and $y^* < x^*$ imply $x^* = y^*$. So $(V, <)$ is acyclic. For example in Figure 13 the acyclic set of values is $V = \{v_1, v_2, v_3\}$ with $v_2 < v_1$ and $v_3 < v_1$ and the maximal strongly connected sets can be read from the figure.*

Definition 2.5. *A universal semantic fucntion $\$$, is a mapping or algorithm associating with each argumentation frame (S, R) a non-empty family of subsets $\{E_i | E_i \subsetneq S\}$ which are conflict free (i.e. for each i and each $x, y \in E_i, \neg xRy\}$.*

Such subsets are called coherent views *or* coherent semantics positions*. Note the restriction $E_i \neq S$!*

Example 2.6. *Given (S, R) we can take as coherent positions for example the following:*

1. *All stable extensions or if there are none take $\varnothing$*

2. *All maximal conflict free subsets.*

3. *The ground extension.*

Definition 2.7. *Let $\$$ be a universal semantic function and let $\mathbb{A}$ be a universal value function. Let (S, R) be an argumentation frame.*

Using $\mathbb{A}$ we can turn (S, R) into an

$$\mathcal{A} = (S, R, V, <, \mathbf{e})$$

being a directional value based network.

Let V_0 be the set of all top points of $(V, <)$, i.e. all points $x \in V$ such that $\neg \exists y \in V (x < y)$.

Consider the networks $\{\mathcal{A}_y | y \in V_0\}$. Let for each y, P_y be a coherent semantics position given by $\$$ for $\mathcal{A}_y$. Let P be the function giving for each y this position P_y. Define the network $\mathcal{A}(V_0, P) = (S(V_0, P), R(V_0, P))$ as follows

$$
\begin{aligned}
S(V_0, P) &= (S - \bigcup_{y \in V_0} S_y) \cup \bigcup_{y \in V_0} P_y \\
R(V_0, P) &= R \upharpoonright S(V_0, P).
\end{aligned}
$$

Let $G(V_0, P)$ be the ground extension of $\mathcal{A}(V_0, P)$ and let S' be the set

$$S' = S - (\{G(V_0, P)\} \cup \{y \in S | \exists x \in G(V_0, P) \text{ such that } xRy\}).$$

Then $\mathcal{A}_\mathbb{V} = (S', R \restriction S')$ is called the remainder. We call the remainder $\mathcal{A}_\mathbb{V}$ because we use the top points V_0 of $\mathbb{V} = (V, <)$ to generate it.

Let $S^0 = \bigcup_i P_i \cup G(V_0, P)$. This set is the set of all points which are supposed to be "in" as the result of our process. The remainder is the system left after the process. Call S^0 the partial directional semantical extension of $\mathcal{A}$.

Remark 2.8. *Let $\mathbb{A}$ and $\$$ be given. Consider (S, R) as a starting point argumentation frame and use $\mathbb{A}$ and $\$$ to form the directional $\mathcal{A} = (S, R, \mathbb{V}, \mathbf{e})$ as defined in Definition 2.1. Let $\mathbf{A}_\mathbb{V}$ be the remainder argumentation network as defined in Definition 2.7.*

$\mathcal{A}_\mathbb{V}$ is an ordinary argumentation frame, let us call it (S_1, R_1). The functions $\mathbb{A}$ and $\$$ will give it a value based annotation $(V_1, <_1, \mathbf{e}_1)$, (coming from $\mathbb{A}$) and $\$$ will give for each frame $(S_{1,v}, R_{1,v})$ where

$$
\begin{aligned}
S_{1,v} &= \{x \in S_1 | \mathbf{e}_1(x) = v\} \\
R_{1,v} &= R_1 \restriction S_{1,v}
\end{aligned}
$$

a family $\{E_{1,v}^i\}$ of coherent semantics positions. Call this new system $\mathcal{A}_\mathbb{V}^{\mathbb{A}, \$}$. We are now back in line for the construction of Definition 2.7 to apply. We can thus get the remainder $\mathcal{A}_{\mathbb{V}, \mathbb{V}_1}^{\mathbb{A}, \$}$ and S_1^0, the partial semantical extension of $\mathcal{A}_\mathbb{V}$ where $\mathbb{V}_1 = (V_1, R_1)$.

Note that S_1 has less points than S and so if we iterate the process again, after a finite number of steps it will terminate (assuming S is finite). The directional extension of (S, R) relative to $(\mathbb{A}, \$)$ is defined as the union of all the partial directional semantic extensions we get in the process. The total directional semantics for (S, R) relative to $\mathbb{A}$, $\$$, is the family of all directional extensions which we can get using the process just described.

Remark 2.9. *If we take $\mathbb{A}$ as the universal function of Baroni's maximal strongly connected sets and $\$$ as maximal conflict free subsets, then the resulting directional semantics as described in remark 2.8 for a given (S, R) is the same as the CF2 semantics for (S, R). This follows from the results of [21].*

142

3 Self-fibred and general context semantics with loops

3.1 General discussion

The considerations in the previous two sections dealt with directional context semantics. However, in actual practice, when there is a conflict between systems, the conflict is bidirectional. We need to deal with general values networks $\mathbb{V} = (V, <)$ which are not necessarily acyclic, and whose elements are associated with argumentation networks. This means that in Definition 2.1 we need to ignore the requirement that the graph $\mathbb{V} = (V, <)$ of values is acyclic. We must consider the case where $\mathbb{V}$ is a general binary relation network.

Let us now set the scene for the results of this section. We are given a system $(V, R_<)$, where V is a finite set and $R_<$ is a binary relation on V. We are also given a system (S, R) where S is a finite set and R is a binary relation on S.

We also have a function $\mathbf{f}$ with domain V, such that for every $v \in V, \mathbf{f}(v) = (S_v, R_v)$, where $S_v \subseteq S$ and $R_v = R \upharpoonright S_v$. In other words, $\mathbf{f}(v)$ is a subnetwork of (S, R).

There are two ways of looking at the overall system $\mathbf{F} = S, R, VR_<, \mathbf{f})$.

View 1: F is a fibred argumentation system. $(V, R_<)$ is the meta-level argumentation network and the function $\mathbf{f}$ associates with each $v \in V$, an object level argumentation network $\mathbf{f}(v)$. The network (S, R) is basically another mother super network containing the union of all the $\mathbf{f}(v), v \in V$. Thus Frame 3 is not ignored and considered "out". A natural condition to impose is the equality:

$$S = \bigcup_{v \in V} S_v, R_< = \bigcup_{v \in V} R_v.$$

According to this view, we seek a definition of what it is to be a complete extension of $\mathbf{F}$. Compare this with my paper [3] on fibring networks and consider this view as generalising the research in [3]. The fibring in this view can be iterated. We can define

$$\mathbf{f}_0 = (S, R)$$
$$\mathbf{f}_1 = (S, R, V^1, R_<^1, \mathbf{f}_1)$$
$$\mathbf{F}_{n+1} = (S, R, V^n, R_<^n, \mathbf{f}_n)$$

where $\mathbf{f}_n$ associates with every $v \in V^n$ a level $m(m \leq n)$ network $\mathbf{F}_v$.

We shall explain this view in more detail in Remark 3.1.

View 2: F is a general value based argumentation network in the sense of Definition 2.1. To present $\mathbf{F}$ as such we need to define $<$ and $\mathbf{e}$, as required by Definition 2.1. Let $v_1 < v_2$ be defined as $v_2 R_< v_1$.

143

e can easily be defined by

$\mathbf{e}(x)$ for $x \in S$ is the unique $v \in V$ such that $x \in S_v$.

For the above view to be mathematically correct we need to assume the following properties on $\mathbf{F}$.

(F1)   For any $x \in S$, there exists a unique $v \in V$ such that $x \in S_v$. Equivalently we can require that

$$v_1 \neq v_2 \Rightarrow S_{v_1} \cap S_{v_2} = \varnothing$$

(F2)   For $x, y \in S$ we have

$$xRy \Rightarrow \mathbf{e}(x) \leqq \mathbf{e}(y).$$

(F3)

$$S = \bigcup_{v \in V} S_v. \qquad R = \bigcup_{v \in V} R_v$$

Remark 3.1. *Note that the above mentioned two views are conceptually different. View 1, the fibring view, is really a substitution view. Recall from proof theory the cut rule:*

$$\frac{\Delta, A \vdash B \qquad \Gamma \vdash A}{\Delta, \Gamma \vdash B.}$$

Δ, Γ are logical theories and A, B are formulas. Δ and A can prove B and A can be proved from Γ. So we substitute Γ for A and from Δ, Γ we can get B. We first prove A from Γ and then, when we have A, we combine it with Δ and get B. Notice that there is a direction here, first use Γ then use Δ and A.

In argumentation context the cut rule has the same form. Consider an argumentation network (S_1, R_1) where $c \in S_1$.

We need to explain what we mean by $(S_1, R_1) \vdash c$. Let us for simplicity understand that $(S_1, R_1) \vdash c$ means that c belongs to the ground extension of (S_1, R_1).[4]

The cut rule for argumentation becomes the following:
Let E_1 be the ground extension of (S_1, R_1) containing $c \in E_1$. Suppose we have another argumentation network (S_2, R_2) also containing c and let E_2 be the ground extension of (S_2, R_2) with $c \in E_2$. The cut rule for argumentation should say that we can now substitute (S_2, R_2) in place of c in (S_1, R_1) and hope to get the ground extension $E_1 \cup E_2$ for $(S, R) = (S_1 \cup S_2, R_1 \cup R_2)$.

Figure 15 schematically illustrates the substitution.

---

[4]We are giving a simple reasonable definition here in order to explain fibring networks. A proper definition of the notion of $(S, R) \vdash a$ requires detailed analysis.

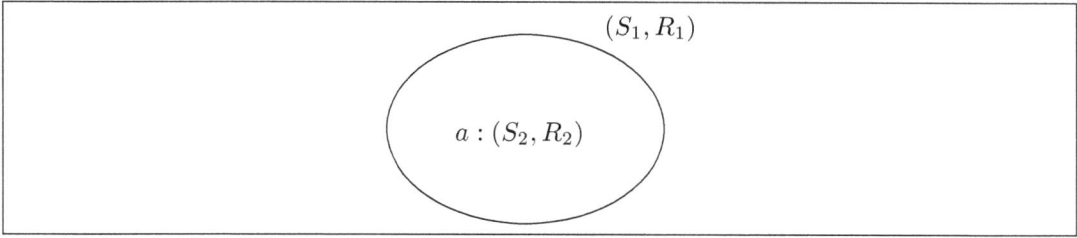

Figure 15:

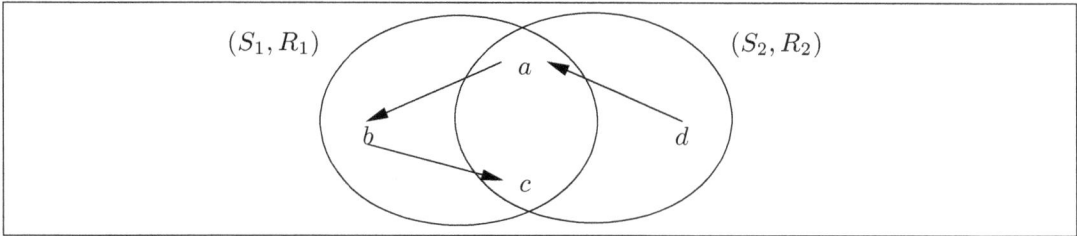

Figure 16:

The substitution becomes more delicate and problematic in the case where $S_1 \cap S_2 \supsetneq \{c\}$, i.e. they have more elements in common besides c. In such a case we have sideways iterations!

See, for example, Figure 16. In this figure (S_1, R_1) contains $\{a, b, c\}$ and S_2 contains $\{a, c, d\}$. However, when we put them together as $(S, R) = (S \cup S', R \cup R')$ we get that d interacts with a. This is a general proof theory problem and does not only arise in our case. Take $\Delta \vdash A$ in linear logic. A must be proved from Δ using each element of Δ exactly once. Suppose now we modify the logic and accept $\Delta \vdash_1 A$ if A is proved using all elements of Δ except possibly one element. Then we have $a, c \vdash_1 a, d, c \vdash_1 c$ but $a, d, c \nvdash_1 c$.

Note that Figure 16 looks just like a values network with loops. We have

$$(V, <) = (\{v_1, v_2\}, \{(v_1, v_2), (v_2, v_1)\})$$

with

$$\mathbf{f}(v_1) = (S_1, R_1)$$
$$\mathbf{f}(v_2) = (S_2, R_2)$$

and

$$(S, R) = (S_1 \cup S_2, R_1 \cup R_2)$$

The above considerations suggest the following general question which we need to address.

145

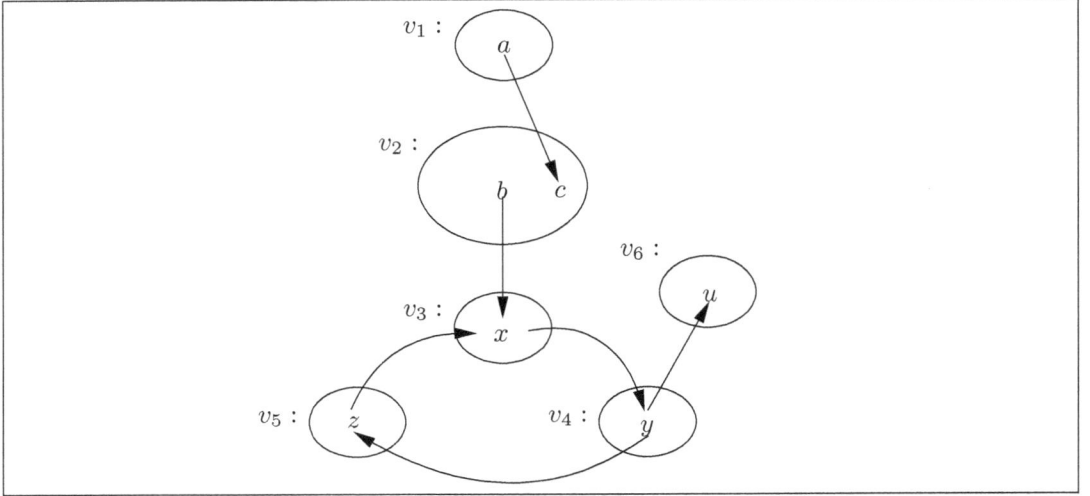

Figure 17:

General fibring question

Given $\mathcal{A} = (S, R)$ and several $\mathcal{A}_i = (S_i, R_i)$ with $S_i \subseteq S, R_i = R \restriction S_i$, define $V = \{\mathcal{A}_i\}$ and define $\mathcal{A}_i < \mathcal{A}_j$ iff for some $x \in S_i$ and $y \in S_j$ we have $(y, x) \in R$; we ask how do we define complete extensions for this system?

3.2 Detailed examples

Let us proceed with several more examples with the hope that we get an idea for a solution.

Example 3.2. *Consider Figure 17. Here we have $(V, <)$ with $V = \{v_1, v_2, v_3, v_4, v_5, v_6\}$* ∎
with $v_2 < v_1, v_3 < v_2, v_4 < v_3$ $v_5 < v_4$, and $v_6 < v_4$.

We have $S = \{u, a, b, c, x, y, z\}$. $R = \{(a, c), (b, x), (x, y), (y, z), (z, x), (y, u)\}$. We have $\mathbf{e}(a) = v_1, \mathbf{e}(b) = \mathbf{e}(c) = v_2, \mathbf{e}(x) = v_3, \mathbf{e}(y) = v_4, \mathbf{e}(z) = v_5$ and $\mathbf{e}(v_6) = u$.

On the face of it, $(V, <)$ has the odd cycle $\{v_3, v_4, v_5\}$. If we regard $(V, <)$ as an argumentation network $(V, R_<)$, with $vR_<w$ defined as $w < v$, the odd cycle $\{v_3, v_4, v_5\}$ is not resolved in the ground extension. So we may think that we have a problem with a cycle but actually we do not have a problem because we have that b breaks the cycle and we actually get the extension is $\{a, b, y\}$. This is because of the way we calculate the overall extension, namely we do not take the frame of v_2 out but only attack its nodes and the surviving nodes from frame v_2 can continue to attack nodes in frame v_3.

The perceptive reader might say that we had no problem with cycles in this case

146

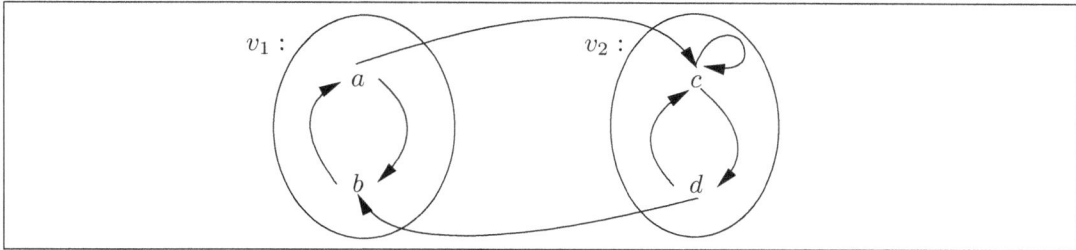

Figure 18:

because (S, R) had no problem with cycles. This is not so. Recall Example 1.2. The directional extension there is different from the (S, R) extension. There is no connection between cycles in $(V, <)$ and cycles in (S, R).

Example 3.3. *Consider Figure 3. It can be considered either as a directional network as in view 2 and its extensions are as calculated in Example 1.1, (namely $\{c, b, \beta\}$ and $\{a\}$) or it can be considered according to view 1 as a fibred network with $V = \{v_1, v_2, v_3, v_4\}$ and $R_< = \{(v_1, v_3), (v_2, v_3), (v_3, v_4)\}$ and $\mathbf{f}(v_i) = $ frame i, for $i = 1, 2, 3, 4$.*

The complete extensions of $(V, R_<)$ are only the unique $\{v_1, v_2, v_4\}$ and common sense dictates that the extensions of Figure 3 should be the unions of the extensions of Frame 1, Frame 2 and Frame 4, which yield $\{a\}$ and $\{c, b\}$.

The difference between the views is more significant when we have loops in $(V, R_<)$. Consider the next example.

Example 3.4. *Consider Figure 18 and let us review our options. In this figure we have a looping $\{v_1, v_2\}$ and mutual attacks between S_{v_1} and S_{v_2}. How do we calculate the complete extensions according to view 1?*

We have the following options:

1. Option view : hierarchical

   We first take an extension of the loop $\{v_1, v_2\}$. There are two extensions $\{v_1\}$ and $\{v_2\}$. Then we take extensions in the surviving frame. From the frame of v_1 we get the extensions $\{a\}$ and $\{b\}$ and $\varnothing$, and from frame of v_2 we get the extensions $\{d\}$ and $\varnothing$. So the totality of the extensions for the fibred complex network is $\varnothing, \{a\}, \{b\}, \{d\}$.

2. Option view 1: directional

   We look at the extension of the loop $\{v_1, v_2\}$ and regard them as giving a direction only and not as "killing" the "out" elements. So the extension $\{v_1\}$

147

*says "start your attacks with frame v_1". So frame v_1 has three extensions $\{a\}$
and $\{b\}$ and $\varnothing$.*

$(v_1 a)$: *starting with a as "in", we get b is "out", c is "out" and d is "in" and
b is again "out".*

*This yields the extension a = "in", b = "out", c = "out", d = "in" for the total
system.*

$(v_1 b)$: *Starting with b as "in" we get a is "out". Continuing with the frame of
v_2 we have the extension $\varnothing$ (all undecided) or $\{d\}$. But d attacks b, so in the
first case we get a = "out" and b, c, d are all undecided, (d is undecided and
so b becomes undecided being attacked by d. a, however, was initially "out"
because b was "in". We do not continue internally in frame v_1 to modify the
value of a to undecided. We stop here!).*

*In the case of the extension $\{d\}$, we get b = "out", and again we stop here and
get the extension d = "in", a = b = c = "out".*

$(v_1 c)$: *Starting with both a = b = undecided, we get two extensions in frame
v_2, $\{d\}$ and $\varnothing$. So the totality of extensions for the overall system is all undecided and d = "in", b = "out" (being attacked by d), a = c = undecided*

(v_2) : *If we take $\{v_2\}$ as indicating direction, we have two extensions for frame
v_2. $\{d\}$ and all undecided. In the first case we get the overall extension d =
"in", b = "out", a = "in", c = "out". In the second case we get overall undecided.*

3. Option view 1: union
   *This option takes the union of the loop as one network. We thus get the overall
   extensions d = a = "in", a = c = "out" and the extension all undecided.*

*Let us now consider our options for view 2, the directional view. In this view we need
direction for the loop $\{v_1, v_2\}$. So our options can be either the directional options
of view 1 or the union option. We believe the union option is the only reasonable
one, given the philosophy behind the approach. The following is how we implement
view 2.*

4. Option view 2: union
   *Given a maximal loop $\lambda = \{v_1, v_2, \ldots\}$ create a new node v_λ to represent and
   replace the loop, and modify $\mathbf{e}$ to $\mathbf{e}^*$.*

$$\mathbf{e}^*(x) = \begin{cases} \mathbf{e}(x), & \text{if } \mathbf{e}(x) \text{ is not part of a loop} \\ v_\lambda & \text{if } \mathbf{e}(x) \text{ is part of a maximal loop } \lambda \end{cases}$$

148

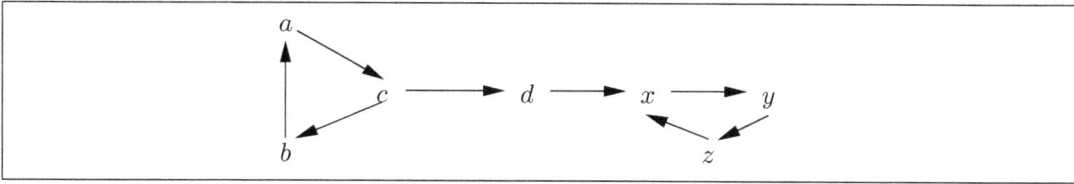

Figure 19: The Network $(V, <)$

Thus we replace $(V, <)$ by $(V^*, <^*)$, where

$$V^* = (V - \{y | y \text{ in some maximal loop } \lambda\})$$
$$\cup \{v_\lambda | \lambda \text{ maximal loop}\}.$$

$$x <^* y \text{ if} \begin{cases} x < y \text{ and } x, y \in V \cap V^* \\ or \\ y = v_\lambda \text{ for some } \lambda \\ and \\ x < z \text{ for some } z \in \lambda \\ or \\ x = v_\lambda \text{ for some } \lambda \text{ and } z < y \\ \quad \text{for some } z \in \lambda \end{cases}$$

The previous example dealt with even loops. It would be helpful to look at an example containing odd loops, and illustrate the use of loop busting methods as described in [21].

We shall use a directional method for getting all CF2 extensions. The best way to explain it is to do a detailed example:[5]

Example 3.5. *Consider Figure 19.*

This figure defines a network $\mathbb{V} = (V, <)$, where $x \to y$ means $y < x$. Here $V = \{a, b, c, d, x, y, z\}$.

We are not giving you any (S, R) nor the function $\mathbf{f}$. Our concern at this stage is only in dealing with the cycles in $(V, <)$.

The CF2 semantics will first bust the top cycle $\{a, b, c\}$ by taking maximal conflict free subsets of the cycle. There are three possibilities, $\{a\}, \{b\}$ and $\{c\}$. For each possibility we get an intermediate network, a subnetwork of $(V, <)$.

See Figures 20, 21 and 22.

---

[5]The way we bust loops in $\mathbb{V} = (V, <)$ is conceptually marginal to the general problem of defining extensions for general fibred or directional networks. The real question is do we deal with

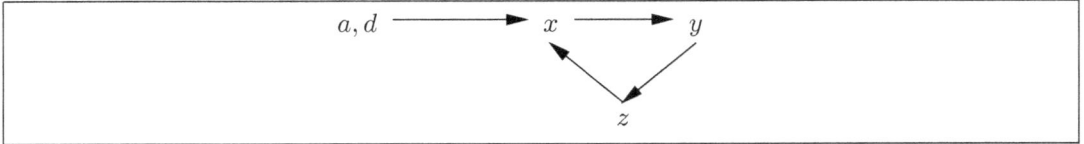

Figure 20:

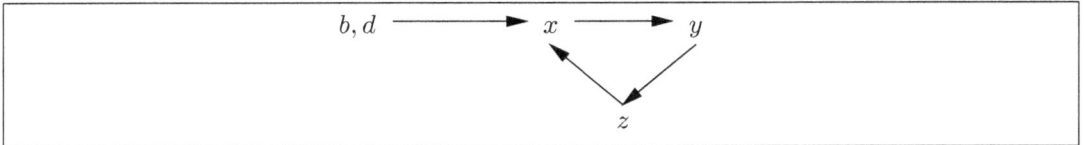

Figure 21:

The theorems in [21] show that the above figures 20, 21 and 22 can be obtained by adding "annihilator" external points (α, β, γ) to the original figure 19, attacking the complements of the maximal conflict free subsets of the cycle $\{a, b, c\}$. This we see in the respective figures 23, 24, and 25.

Continuing to calculate extensions we get the extension $E_1 = \{a, d, y\}$ for Figure 20. For the "equivalent" Figure 23 we get the extension $E_1' = \{\alpha, a, d, y\}$. We say "equivalent" because $E_1 = V \cap E_1'$ (i.e. we get E_1 if we ignore α). Similarly we get $E_2 = \{b, d, y\}$ and the "equivalent" $E_2' = \{\beta, b, d, y\}$.

We do not get a $\{0, 1\}$ extension in the third case, of Figures 22 or the "equivalent" Figure 25 because the loop $\{x, y, z\}$ remains. So we need to bust this loop and CF2 will again take maximal conflcit free subsets of $\{x, y, z\}$ and we get respectively Figures 26, 27 and and 28.

The corresponding respective "equivalent" figures are 29, 30 and 31

Let us see where we are now. We started with the figure with loops, Figure 19. This figure is for $\mathbb{V} = (V, <)$. We ended up with the following figures, which are supposed to give us (eventually, but not yet) loop free $\mathbb{V}_i^*$. The figures are

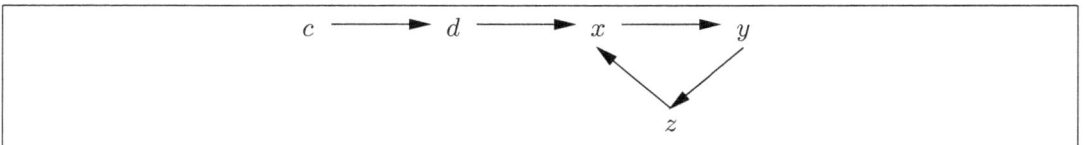

Figure 22:

$\mathbb{V}$ first or do we deal with each $\mathbf{f}(v) = (S_v, R_v)$ first, an how we combine them. However, we want to use this example in the next section which deals with representations of the CF2 semantics, and so we give great details in this example to how we use CF2.

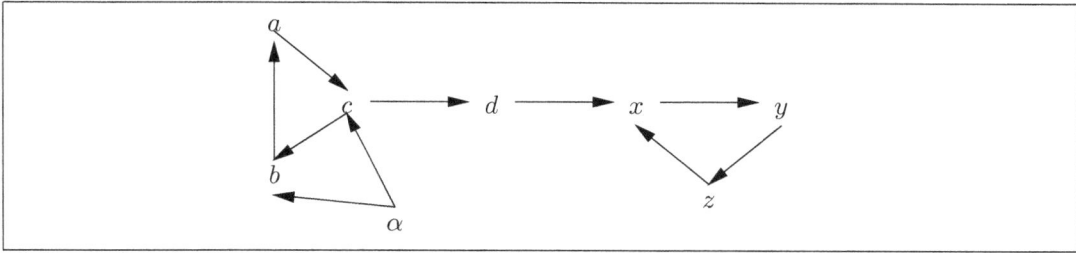

Figure 23:

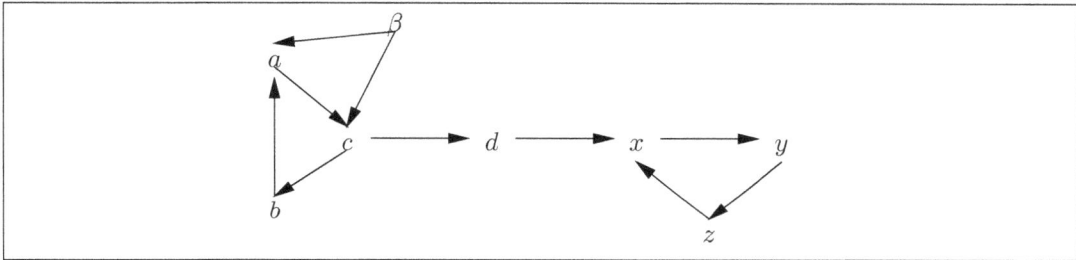

Figure 24:

1. *Figure 23, yielding extension E_1*

2. *Figure 24, yielding extension E_2*

3. *Figure 25, yielding extension E_3*

4. *Figure 29, yielding extension E_4*

5. *Figure 30, yielding extension E_5*

These figures still have loops in them geometrically, but the loops are busted by the external nodes $\alpha, \beta, \gamma, \xi, \eta, \zeta$ acting as "annihilators", which destroy the loops. So we have for each network of each figure a unique $\{0, 1\}$ ground stable extension.

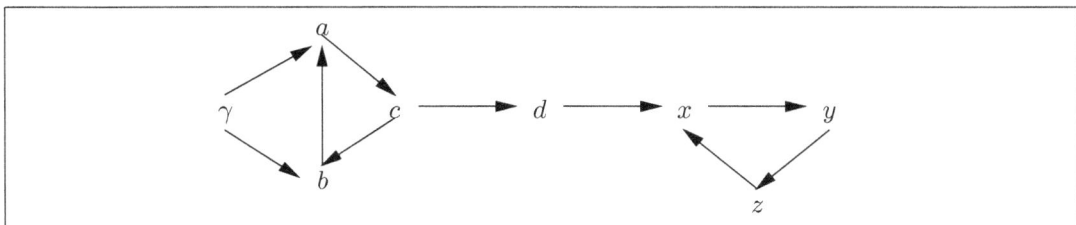

Figure 25:

$$c \to d \to x$$

Figure 26:

$$c \to d \qquad y$$

Figure 27:

We need one more step before we are done. Look again at Figure 30. It has two external annihilators γ and ζ. Really we need only one annihilator, say δ_5 to do their job. In principle in each figure case only one external annihilator is needed. So our final figure is Figure 32

We similarly get the rest of the figures. Figures 33, 34, 35, and Figure 36

The five Figures, 32, 33, 34, 35, and 36 can be combined into a single figure 37, where the set $\{\delta_1, \ldots, \delta_5\}$ is a total loop with δ_i attacking δ_j for $i \neq j$. Any extension of this loop will choose one and only one $\{\delta_i\}$ and the remaining nodes will behave like Figures 32, 33, 34, 35, and 36, respectively.

Remark 3.6. *This remark is for the sake of the next section, where we deal with representation theorem for CF2 semantics. Let us recall the discussion starting with Figure 19. This figure has two odd loops in and for whatever reasons, we wanted to bust these loops and see what extensions we can get.*

We used the CF2 semantics of Baroni et al. [1] using a directional algorithm with annihilators proved equivalent to the definition in [1] in our paper [14]. We calculated all the CF2 extensions of Figure 19 and ended up with a single Figure 37, representing a network containng additional nodes, whose traditional Dung complete extensions generated the CF2 extensions of the (network of) original figure.

This procedure is general and suggests the following theorem:

$$c \to d \qquad z$$

Figure 28:

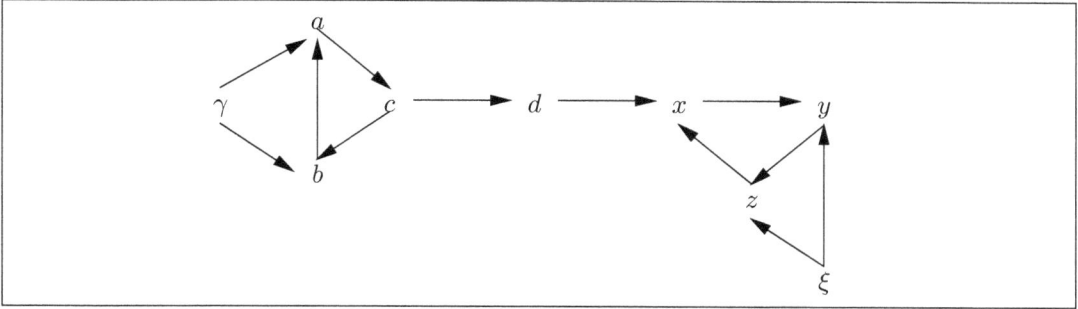

Figure 29:

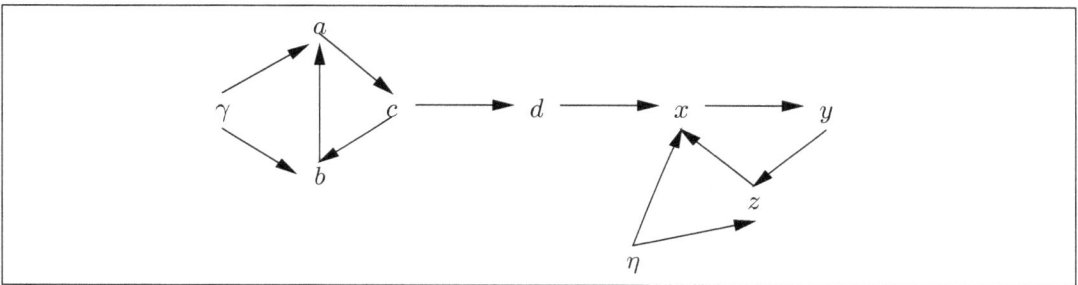

Figure 30:

Theorem 3.7. *Given a finite argumentation network* (S, R). *Then there exists a finite argumentation network* (S', R') *with* $S \subseteq S'$ *and* $R = R' \cap S \times S$ *such that the following holds:*

1. *If* E *is a CF2 extension of* (S, R) *then there exists a complete traditional Dung extension* E' *of* (S', R') *such that* $E = S \cap E'$.

2. *If* E' *is a complete Dung extension of* (S', R') *then* $E = S \cap E'$ *is a CF2 extension of* (S, R).

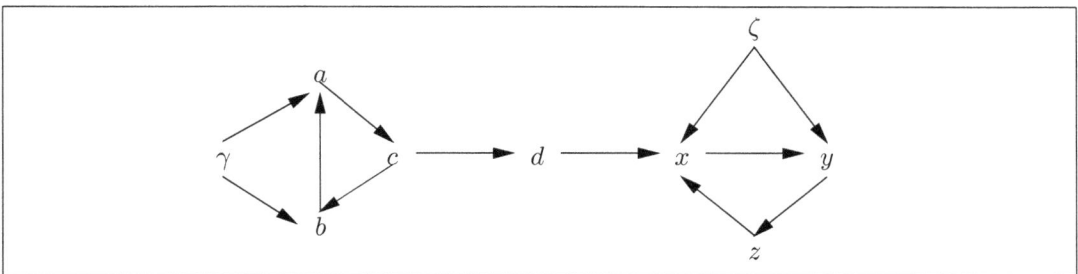

Figure 31:

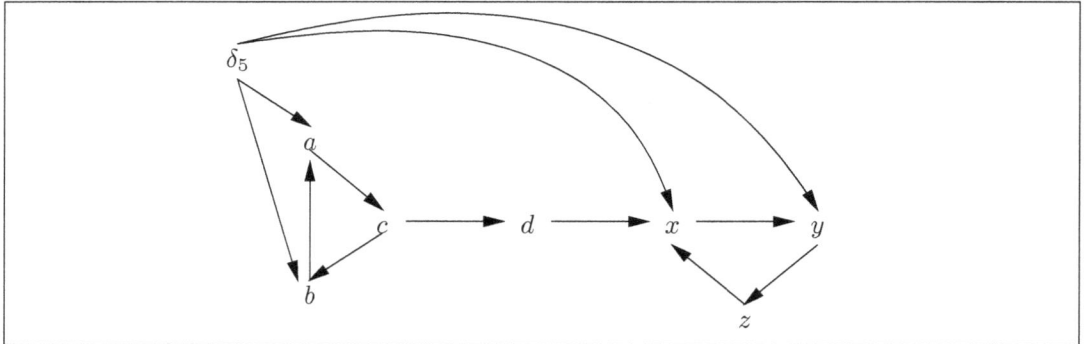

Figure 32:

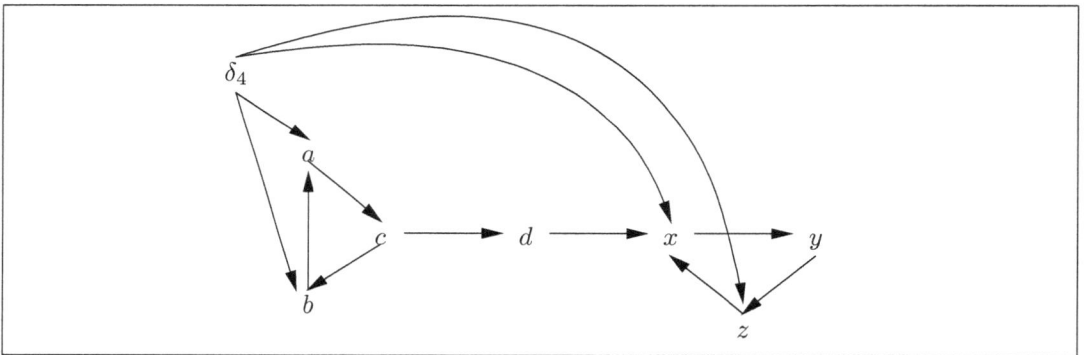

Figure 33:

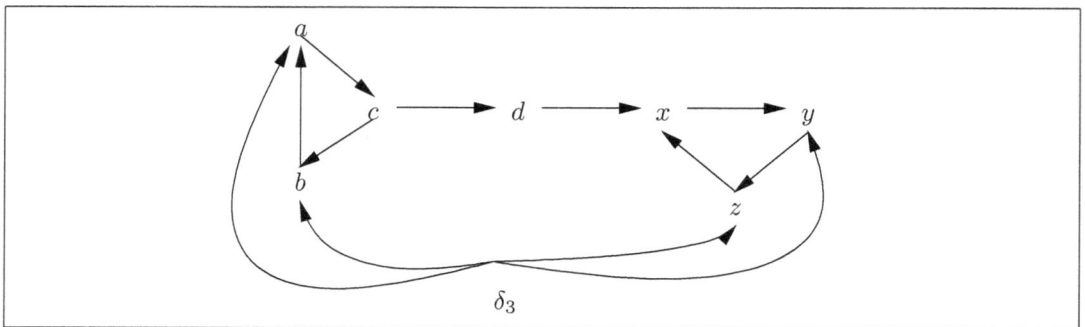

Figure 34:

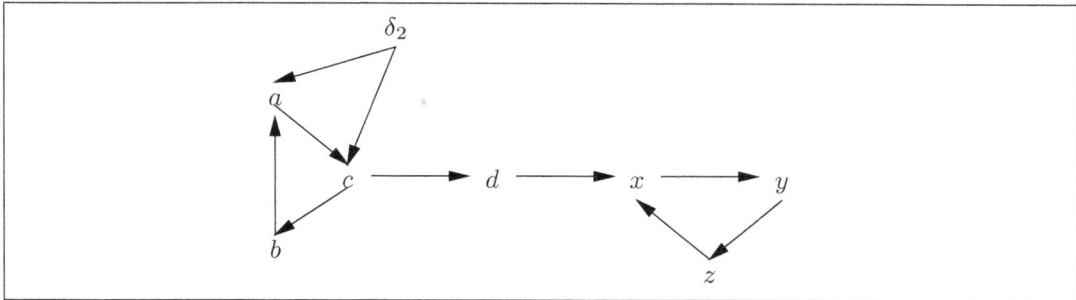

Figure 35:

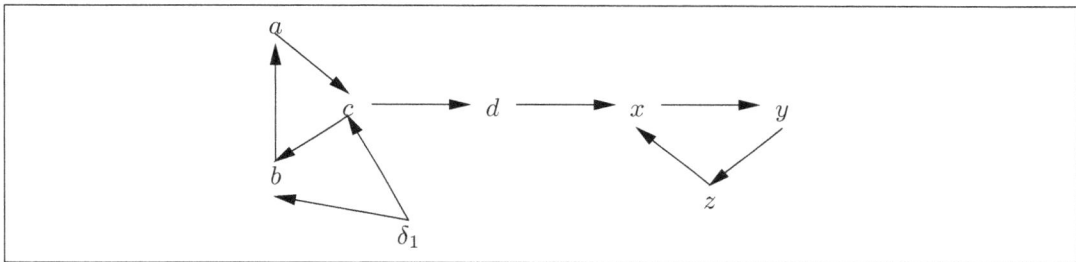

Figure 36:

Proof. Suggested by the construction in Example 3.5. □

The perceptive reader may be skeptical about this result. He may say that whenever we have (S, R) with any extensions $E_1, E_2, E_3, \ldots$. We can use a brute force approach and for each E_i annihilate its complement $S - E_i$ by a δ_i and then create a single network with $S \cup \{\delta_i\}$ with suitable attacks to achieve the theorem. My answer is that this is indeed true but if we do that we lose all the information of the attack structure of (S, R). We can equally eliminate R altogether and just look at S and its subsets E_i.

Our annihilators actually follow the CF2 directional sequence for calculating the extensions and just help in loop busting. We need still to follow the attack structure in $R = R' \cap (S \times S)$ to calculate the extensions. Having said all that, we are going to do better in Section 4. We keep S fixed and change R to get our representation theorem.

3.3 Technical definitions and summary

Following the examples and discussions in the previous subsections, we are now ready to summarise and give final technical definitions and remarks.

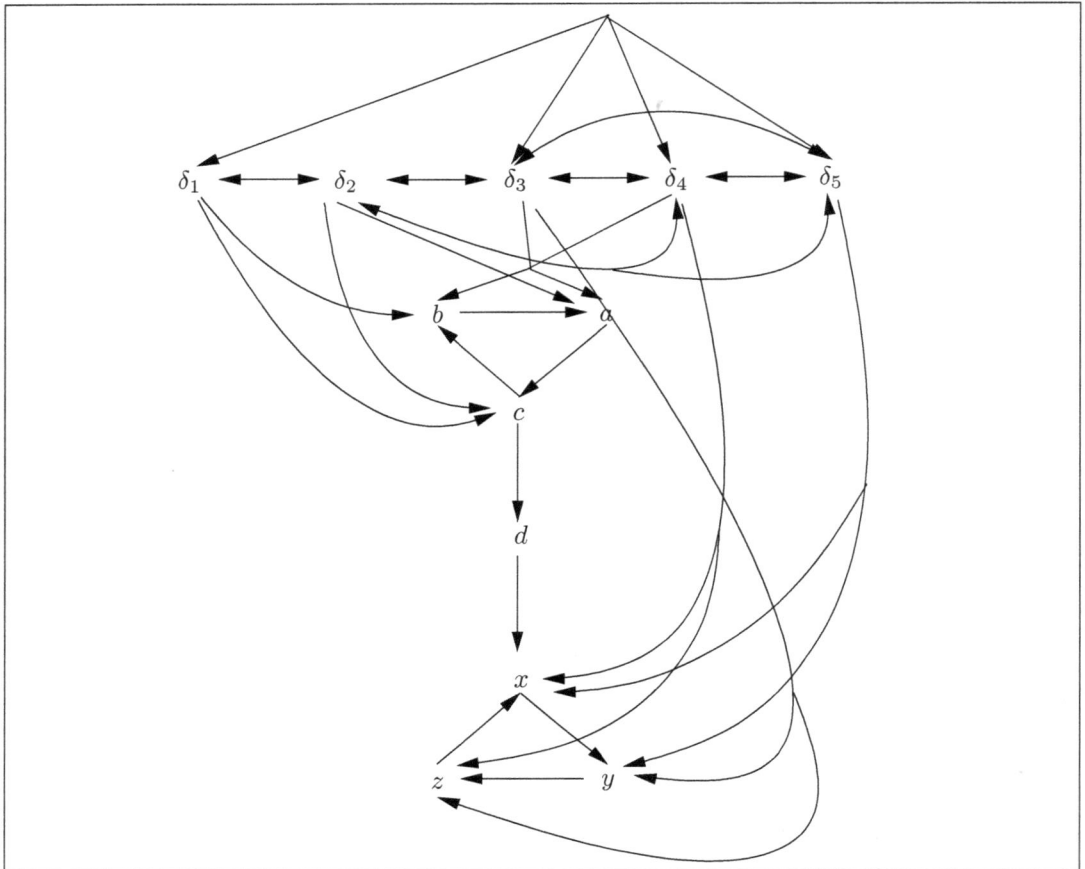

Figure 37:

Definition 3.8. *1. Let (T, ρ, Λ) be a finite tree with root Λ and tree relation ρ. $x\rho y$ means that x is the unique predecessor of y. Being a tree means that each $y \neq \Lambda$ in the tree has a unique sequence $(y_0 = \Lambda, y_1, \ldots, y_k = y)$ such that for each $0 \leq i < k$ we have $y_i \rho y_{i+1}$.*

2. We divide T into levels. Let level 0 be $V_0 = \{\Lambda\}$. Let level $n + 1$ be

$$V_{n+1} = \{y| \text{ for some } x \text{ of level } n \text{ we have } x\rho y\}.$$

3. For $y \in T$, let T_y be the subtree of all points z in T such that $y\rho^* z$, where ρ^* is the reflexive and transitive closuer of ρ. T_y is the set of all points in T which are "above" y, including y itself (trees go "up" with ρ).

156

4. Let $\mathbf{f}$ be a function associating a finite argumentation network (S_y, R_y) with every $y \in T$, i.e. $\mathbf{f}(y) = (S_y, R_y)$. The following is assumed to hold for all $x \in T$.

$$(S_x, R_x) = (\bigcup_{x \rho y} S_y, \bigcup_{x \rho y} R_y)$$

5. Let $x \in T$ and let $\rho(x)$ be $\{y | x \rho y\}$. Let $<_x$ be a binary relation on $\rho(x)$. We can assume that

$$x \neq y \rightarrow \rho(x) \cap \rho(y) = \varnothing.$$

6. We call the system

$$\mathbf{F} = (T, \rho, \mathbf{f}, <_x, x \in T)$$

a fibred argumentation system.

Example 3.9. *1. Recall Definition 2.1. We defined a general value based argumentation system $(S, R, V, <, \mathbf{e})$. This definition is a special case of our current definition 3.9 as follows: Assume $\Lambda \notin V$:*

(a) Let $T = \{\Lambda\} \cup V$

(b) Let $\Lambda \rho v$ hold for $v \in V$

(c) Let $\mathbf{f}(\Lambda) = (S, R)$
Let $\mathbf{f}(v) = (S_v, R_v)$, where
$S_v = \{s \in S | \mathbf{e}(v) = s\}$
$R_v = R \upharpoonright S_v)$.

(d) Let $<_\Lambda = <$

We need to assume that $S = \bigcup_{v \in V} S_v$ to fit Definition 3.9, but this assumption is natural and could have been added to Definition 2.1.

2. The system of Figure 16 can be described according to our definition as follows:

(a) $T = \{\Lambda, v_1, v_2\}$

(b) $\Lambda \rho v_1, \Lambda \rho v_2$

(c) $\mathbf{f}(v_1) = (S_1, R_1)$
$\mathbf{f}(v_2) = (S_2, R_2)$
$\mathbf{f}(\Lambda) = (S_1 \cup S_2, R_1 \cup R_2)$

(d) $<_\Lambda$ can be defined in two ways. Either as $\{v_1 <_\Lambda v_2\}$ giving v_2 priority of direction, or symmetrically as $\{v_1 <_\Lambda v_2$ and $v_2 <_\Lambda v_1\}$.

Example 3.10. *This is a merging of networks example. Assume we have n agents,*
$v_1, \ldots, v_n$. *Each agent v has his own network* (S_v, R_v) *and possibly his own complete
extension* E_v. *The agents want to work together and merge their networks into a
common* (S_Λ, R_Λ), *where* $S_\Lambda = \bigcup_v S_v$ *and* $R_\Lambda = \bigcup_v R_v$.

The merging problem is how to define extensions E_Λ *on* (S_Λ, R_Λ) *from known
extensions of* $(S_v, R_v), v \in V$.

Our approach allows also for a prioirty ordering $<_\Lambda$ *on the agents. In traditional
merging problems there are no priorities. If we have* $<_\Lambda$, *we can use it in the merge
process.*

Here we have $\Lambda \rho v$, *for all v, but this is only a formality. The widespread approach
to this problem is to use voting. See [27] and the references there. So the priority*
$<_\Lambda$ *can give more weight to some voters.*

Remark 3.11. *The discussion in Examples 3.9 and 3.10, presented the general value
based argumentation system of Definition 2.1 as a level 1 fibred argumentation system
of Definition 3.8. A by-product of this reduction is that we have various ways of
defining extensions for the root* (S_Λ, R_Λ) *of any level 1 fibred argumentation system,
following the options and examples discussed in Sections 3.1 and 3.2. Similarly we
have various algorithms for merging (not presented here, but see [27]). We can use
this to define sematnics for the general networks of Definition 3.8. The point of this
remark is to be able to assume that given a level 1 network* $(\{\Lambda\} \cup V, <, \mathbf{f})$ *with* $< \subseteq$
$V \times V$ *and* $\mathbf{f}(\Lambda) = (S_\Lambda, R_\Lambda) = (\bigcup_{v \in V} S_v, \bigcup_{v \in V} R_v)$ *we can have a choice of algorithms
(say* $\mathbf{Alg}_{\text{base}}$ *and* $\mathbf{Alg}_{\text{top}}$*) such that* $\mathbf{Alg}_{\text{base}}$ *can give all* $\mathbf{Alg}_{\text{base}}$ *extensions of any*
$\mathbf{f}(v) = (S_v, R_v)$ *(for example we can take CF2 semantics for every v, or we can take
different semantics for different* $v \in V$. *The important point is that for each* $v \in V$,
we can generate coherent views of "extensions") and $\mathbf{Alg}_{\text{top}}$ *can use the extensions
of* $\mathbf{Alg}_{\text{base}}$ *to give all possible extensions for* (S_Λ, R_Λ).

*This procedure can be iterated to levels 2, 3, . . . networks as done in the next
definition 3.12.*

Definition 3.12. *Let* $\mathbf{F} = (T, \rho, \mathbf{f}, <_x, x \in T)$ *be a fibred argumentation system.
Let* $\mathbf{Alg}_{\text{base}}$ *and* $\mathbf{Alg}_{\text{top}}$ *be two algorithms for computing all extensions for level 1
networks as discussed in Remark 3.11. We define the notion of all extensions of an
arbitrary* $\mathbf{F}$ *by recursion on the levels in* $\mathbf{F}$.

For $\mathbf{F}$ *of level 0 and 1 we can use* $\mathbf{Alg}_{\text{base}}$ *and* $\mathbf{Alg}_{\text{top}}$.

Let $\mathbf{F}$ *be a system of level* $n + 1$ *and assume that we can calculate all extensions
of* (S_Λ, R_Λ) *for any* $\mathbf{F}'$ *of level* $\leq n$. *Consider* $\mathbf{F}$ *and consider all elements* $y \in \rho(\Lambda)$.
Let $\mathbf{F}_y, y \in \rho(\Lambda)$ *be the subsystem using y as a root, i.e* $\mathbf{F}_y = (T_y, \rho \restriction T_y, \mathbf{f} \restriction T_y, <_x
, x \in T_y)$.

By the induction hypothesis we have an algorithm $\mathbf{Alg}_y$ giving all extensions of $\mathbf{f}(y) = (S_y, R_y)$. This algorithm can serve as our basis for defining a new $\mathbf{Alg}_{\text{base}}$. This new $\mathbf{Alg}_{\text{base}}$ will use on each (S_y, R_y) the algorithm $\mathbf{Alg}_y$ to generate its extensions

We now look at the level 1 system

$$(T_1 = \{\Lambda\} \cup \rho(\Lambda), <_\Lambda, \mathbf{f} \restriction T_1).$$

We know how to find all extensions using the pair ($\mathbf{Alg}_{\text{top}}$, new $\mathbf{Alg}_{\text{base}}$). The extensions we thus get are the extensions for our $\mathbf{F}$ of level $n + 1$.

4 The adjustment approach to CF2 semantics, an orientation

Our paper [21] offered several methods of handling loops in networks (S, R), which were metalevel. They did not change the attack relation R. The methods basically tried to identify extensions in various ways, using various algorithms and to the extent that we moved to another network, it was to show soundness; to explain what the algorithm was doing in terms of another network. In the previous section in Example 3.5, we showed in great detail how the CF2 semantics for breaking loops, can be obtained for the network of Figure 19. Remark 3.6 and Theorem 3.7 showed how to reduce the question of dealing with loops and finding the CF2 extensions for (S, R) to finding traditional Dung extensions for another network (S', R'). We indicated that this reduction was somewhat unsatisfactory. This section offers another method (the adjustment method) which keeps S and changes (adjusts) only the attack relation R. The adjustment method actually says that since (S, R) has problematic loops, let us consider another network $(S, \check{R})$ instead. This is a use of the idea of intertranslatibility of [18, 19]. This approach might still be open to criticism because it may seem that we are evading the problem, rather than solving it.

The perceptive reader might say that we are not dealing with (S, R) but with a different network $(S, \check{R})$! Whereas [21] tried to compute the (S, R) extensions and when presented with a loop tried to fix the algorithm and get results, the adjustment method abandons (S, R) and goes to $(S, \check{R})$ instead. The perceptive reader might think that this method is methodologically faulty.

The answer is that it all depends on how near $(S, \check{R})$ is to (S, R). To take a family example, if I offer you soup and you don't like it and I give you a sandwich instead, then I have abandoned the soup in favour of a sandwich. However, if I put

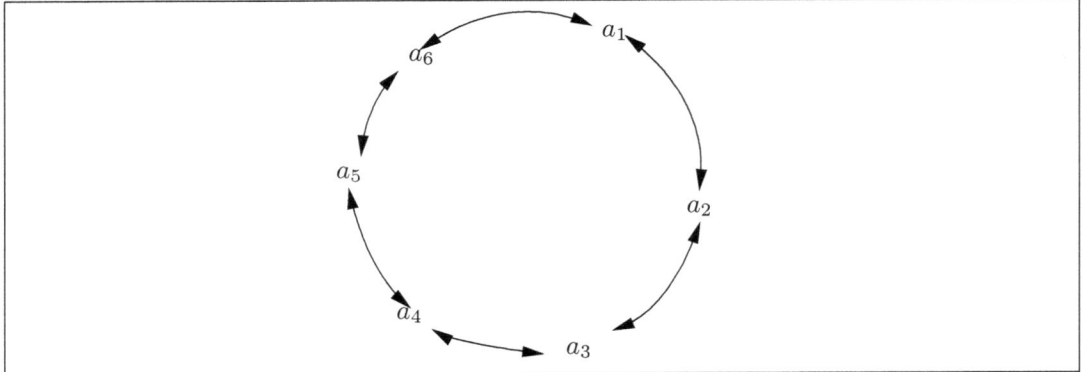

Figure 38:

pepper in the soup and thus make it nicer and more acceptable to you, then I have fixed it. Let us illustrate this idea by an example which leads to a nice theorem.

Example 4.1. *Consider a 6-cycle (S, R), where $S = \{a_1, \ldots, a_6\}$ and with $a_1 R a_2, \ldots, a_6 R a_1$. This cycle has the traditional non-empty extensions $\{a_1, a_3, a_5\}$ and $\{a_2, a_4, a_6\}$. (It also has the all undecided empty extension.) It also has the CF2 extensions $\{a_1, a_4\}, \{a_2, a_5\}, \{a_3, a_6\}$.*

Let us focus our attention on the CF2 approach. Let us change R into $\check{R}$, being its symmetric closure. We have

$$x \check{R} y \text{ iff } x R y \vee y R x$$

We get Figure 38

It is easy to see that for any strongly connected network (S, R) the maximal conflict free subsets of (S, R) and the maximal conflict free subsets of its symmetric closure $(S, \check{R})$ are the same. So they both have the same CF2 extensions. There is, however a fundamental difference between the way we compute the extensions for (S, R) and $(S, \check{R})$. In (S, R) we do the CF2 approach, not the traditional Dung style approach. For $(S, \check{R})$, on the other hand, we can take all the traditional Dung style non-empty extensions!

Thus the CF2 semantics becomes (at least for the case of strongly connected sets) an adjustment approach semantics.

Start with (S, R). Move to $(S, \check{R})$, take all non-empty extensions for $(S, \check{R})$ and these will be your CF2 extensions for (S, R), (it is important to note here that we do not include the empty extension of $(S, \check{R})$).

It seems this may lead us to some sort of a general theorem. To find out we need more examples.

We have to be careful here at the boundaries, i.e. when E is empty. Take (S, R) to be $(\{a\}, \{((a, a)\})$, i.e. a single point attacking itself. The symmetric closure is (S, R) itself (i.e. $(S, \breve{R}) = (S, R)$). Thus there are no non-empty extensions to $(S, \breve{R})$, but (S, R) does have the empty CF2 extension and so the CF2 and traditional extensions do not match. So we had better talk only about (S, R) not containing self attacking elements!

We can prove that if we delete from a network (S, R) all the self attacking elements to obtain (S', R'), then the CF2 extensions of (S', R') (which do not contain self attacking elements anyway) and the non-empty Dung complete extensions of $(S', \breve{R}')$, remain the same.

We are aiming at something like Conjecture 4.2.

Conjecture 4.2. *There exists an algorithm which gives us for any (S, R) without self attacking elements, another network (S, R_1) such that for any non-empty subset $E \subseteq S$, (1) and (2) are equivalent:*

1. *E is a CF2 extension in (S, R)*

2. *E is an ordinary traditional Dung complete extension in (S, R_1).*

This conjecture is certainly true if (S, R) is a strongly connected set. So maybe we can take R_1 to be the result of adding to R all the symmetric closures of all maximal strongly connected subsets in (S, R).

Example 4.3. *Consider Figure 39. The top maximal strongly connected set is $\{\alpha, \beta, \gamma\}$ and the bottom one is $\{x, y, z\}$. Our conjecture says that we should take the symmetric closure of each cycle and get Figure 40 and that on this new figure the traditional extensions would give all CF2 extensions on Figure 39.*

This is not the case!

Let us check Figure 39. We can start by letting (CF2 style) $\beta = $ "in", and $\gamma = \alpha = $ "out". Therefore $x = $ "out" and $y = $ "in" and $z = $ "out".

If we do the same with Figure 40, i.e. start with $\beta = $ "in", we get Figure 41.

The CF2 extensions of Figure 39 do not allow for $\beta = $ "in" and $z = $ "in", while Figure 41 does allow for that.

By symmetry we would have the same problem if we had started with the extesnsion $\gamma = $ "in", $\alpha = \beta = $ "out".

We ask ourselves, is there another formulation of the conjecture which will be a general theorem? The idea is, after all, very attractive.

We want something like the following:

Let (S, R) be a network. Then there exists an adjustment of R into R_{CF2} (same S) where R_{CF2} may be possibly a higher level attack relation, such that for any set $E \subseteq S, E \neq \varnothing$ we have (1) iff (2)

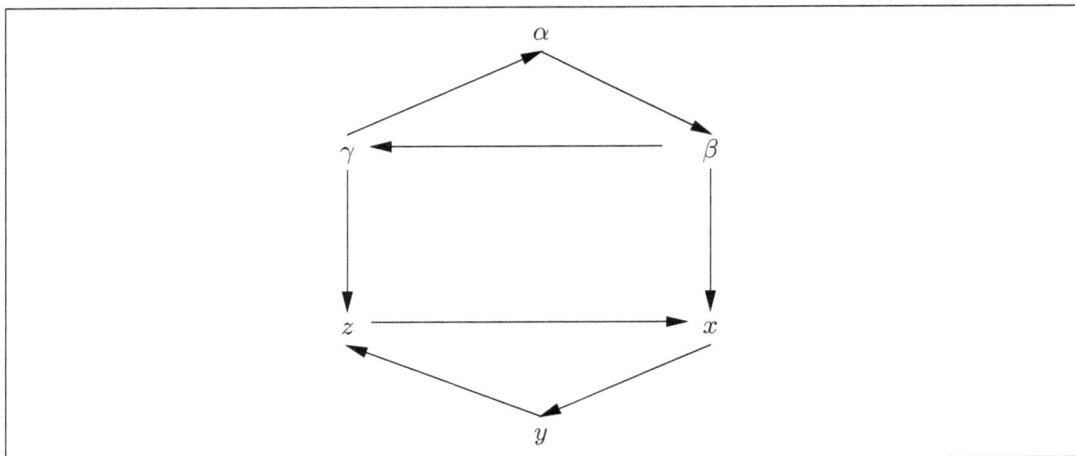

Figure 39:

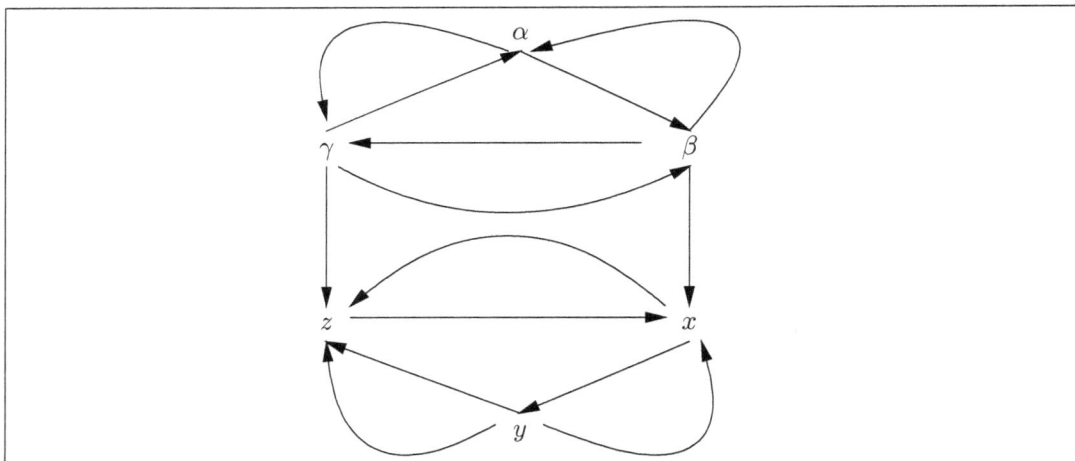

Figure 40:

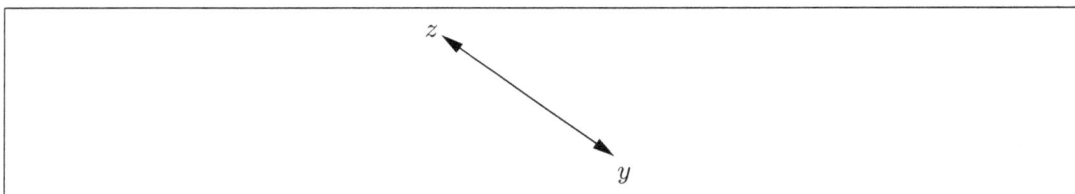

Figure 41:

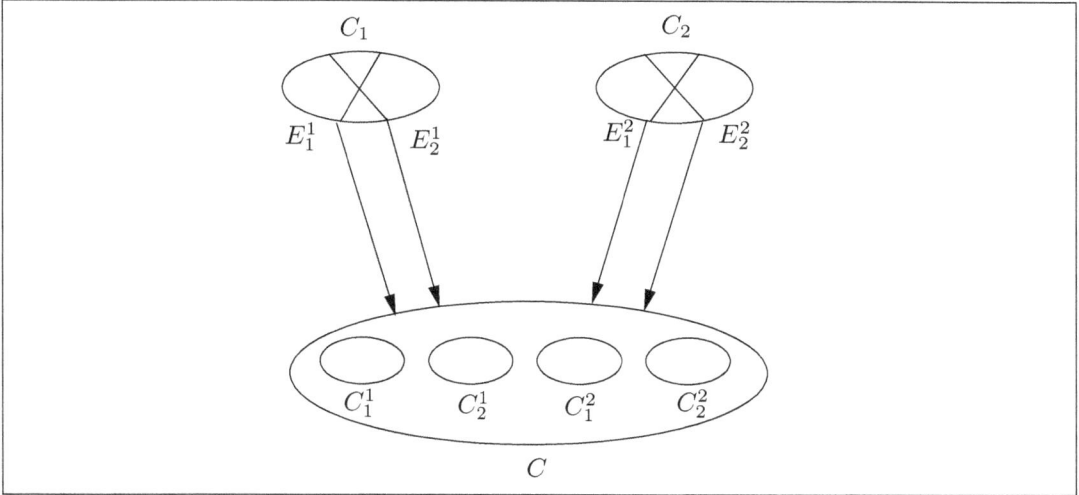

Figure 42:

1. E is a CF2 extension of (S, R).

2. E is a Dung traditional extension of (S, R_{CF2}).[6]

The next example analyses the situation, with a view to formulating a general theorem.

Example 4.4. *Consider Figure 42. This is a schematic figure of what can happen in our directional computation of CF2 extensions, if we follow the procedure of Remark 2.9.*

C_1 and C_2 are maximal strongly connected sets attacking the maximal strongly connected set C. We are in the middle of the construction of the CF1 extension and we have several possibilities to continue the construction by choosing extensions in C_1 and in C_2. In Figure 42 we show two for each. E_1^1 and E_2^1 and E_1^2 and E_2^2.

Suppose we choose E_1^1 and E_1^2 as the extensions. These attack C and propagate and we are left with the strongly connected subset C_1^1 of C, which we can use to continue our inductive constructions.

Had we chosen a different pair, say E_2^1 and E_2^2, we would have been left with a possibly different strongly connected subset C_2^2. We know that to continue the construction we can take the symmetric closure on C_1^1 and this will work for the current inductive construction which chose the extensions E_1^1 and E_1^2 but it will not

---

[6]We need to show how traditional extensions can be defined for networks with higher level attacks. See [7, 8, 9], and the Appendix.

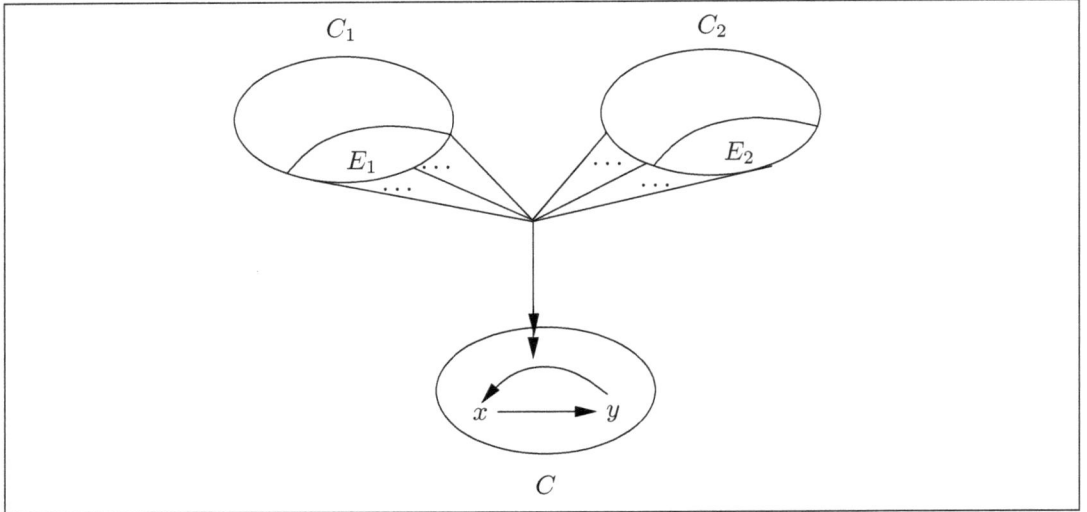

Figure 43:

work had we made a different choice and taken the extensions E_2^1 and E_2^2. So we do not know beforehand what symmetric closure to take of what parts of C. The choice depends on the inductive process itself. More explicitly, let x, y be in C and assume x attacks y. For some choice of extensions, say E_1^1 and E_1^2, after they attack C and propagate, it may be the case that x and y will remain in a loop and so we can add to the loop that y attacks x. However, had our process made a different choice, say E_2^1 and E_2^2, x and y could be out of the loop and so we must not add that y attacks x. So what do we do? We must take a higher level approach. We use higher level joint attacks. See the Appendix for a discussion and examples. Briefly, x and y jointly attack z means that z is "out" exactly if both x and y are "in". See also [3, 24] for joint attacks. Level 1 higher level attacks is when there is an attack on an attack, i.e. an attack on an arrow $x \to y$. If successful the arrow is "out". See [7, 8] for this concept. The appendix is self contained and explains everything needed for this paper.

We use this device. Let E_1 be CF2 extension of C_1 and E_2 of C_2. Take the symmetric closure of C and let x attacks y be at C. Then we have added that y attacks x. It may be that the result of the combined attack of E_1 and E_2 on C requires that y does not attack x. In this case we form a joint attack from E_1 and E_2 on the arrow $y \to x$, to cancel it. Figure 43 shows this situation.

Thus for any CF2 extension E_1 and E_2 we mount joint attacks from E_1 and E_2 onto any arrow $y \to x$ that should not be there (i.e. was added when we took the symmetric closure of C). If the CF2 construction process chooses a different set of

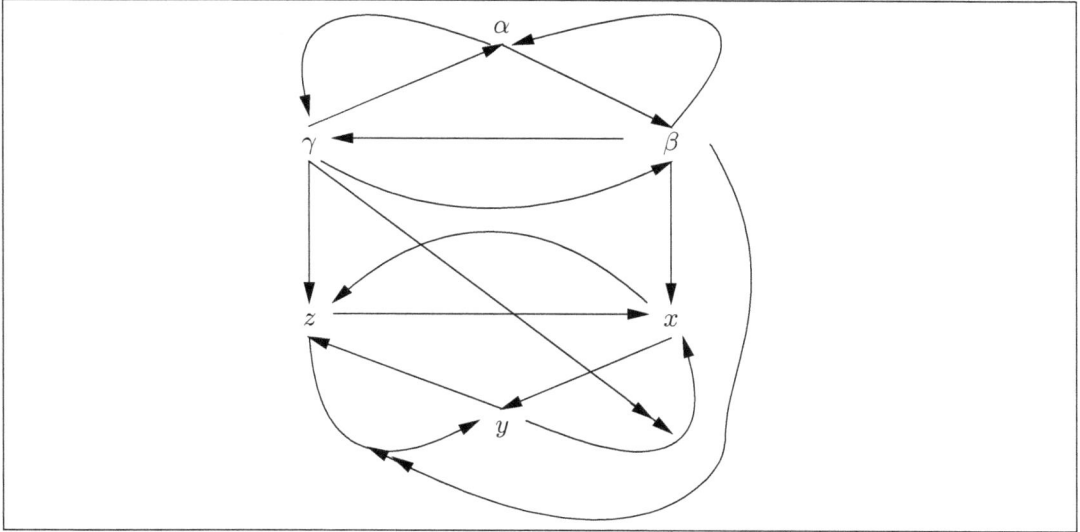

Figure 44:

extensions, say E'_2, E'_2, then these joint attacks will not fire and the joint attacks from E'_1 and E'_2 will fire.

So really the embedded joint attacks in the higher level network ensures that the process of construction proceeds as if we have never added any arrows to the original network (S, R).

Figure 44 shows how to modify Figure 40 in the spirit of what we have just discussed. We add the double arrow attacks from β and from γ.

Our disucssion shows that the following theorem is true.

Theorem 4.5. *Let (S, R) be an argumentation network without self attacking elements. Then there exists a higher level argumentation network (S, R_{CF2}) (same S, different attacks) such that (1) and (2) are equivalent for any subset $E \subseteq S$.*

1. *E is a CF2 extension in (S, R)*

2. *E is a traditional Dung extension in (S, R_{CF2}).*

Proof. As discussed and illustrated in Example 4.4. □

Remark 4.6. *We need to figure out what to do with (S, R) which contains self attacking elements. My guess is that we need to adjust it to a network (S', R') without any self attacking elements and then apply Theorem 4.5.*

165

Theorem 4.7. *Let (S, R) be a network and let D be the subset of all self attacking elements. Then the network with $S - D$ and R restricted to $S - D$ has the same CFD extensions (recall Remark 1.3) as (S, R).*

Proof. By induction of the directional construction of the CF2 semantics as given in [21] and in Remark 2.9. It is sufficient to prove this for strongly connected sets, since the CF2 extensions of such sets drive the induction process. Clearly, if any maximal conflict free subset of S would be maximal in $S - D$ but not in S, then it can be extended in S with at least one more element x. x is not in D because the elements of D are self attacking. So x is in $S - D$ and so E is not maximal in $S - D$. The other direction is even more obvious. □

Remark 4.8. *Note for example that the theorem holds for the network pair of Figure 39 and Figure 44.*

The perceptive reader might raise some objections to Theorem 4.5 as follows:

Question 1: You are using higher level networks. I don't like it.
Answer 1: Higher level networks are well accepted and used by now. They make sense and arise in many contexts. See references [7, 8].

Question 2: To find R_{CF2} you need to follow the CFD inductive construction process and compute all the CF2 extensions of (S, R) in order to define the double arrows of R_{CF2}. So you give us R_{CF2} which yields all extensions only after you have already computed all of them.
Answer 2: We are giving an existence representation theorem of conceptual value. It is not a computational complexity theorem. When you give a representation theorem you assume you know what you have and you want to represent it in a different way. In many cases it is true, that the computation on the representation is faster than on the original (see Remark 4.8, for example) but this would be only a bonus.
Furthermore, the basic idea is to take the symmetric closure of the attack relation. We can use only this idea as follows:
Begin with a network (S, R)

1. Identify the top loop maximal strongly connected sets. Delete all self attacking elements.

2. Make the attack relation symmetric on these sets.

3. Find all traditional extensions on the resulting sets.

4. Propagate the attack in the network for each extension.

166

5. Take out all points with clear $\{0,1\}$ value. Let the residual network be (S_1, R_1).

6. If (S_1, R_1) is empty stop, otherwise go to 1. and proceed with (S_1, R_1).

See also our answer to Question 3.

Question 3: Are there any advantages to your theorem?
Answer 3: Yes, there are. Higher level networks have a simple and natural equational semantics. This is an advantage, as the equations involve only the nodes in S as variables. So, as a result, we have the following theorem.

Theorem 4.9. *There exists an algorithm yielding for any network (S, R) a system of equations $\mathbb{E}$ involving the elements of S as variables, such that all the $\{0,1\}$ solutions of $\mathbb{E}$ give exactly the CF2 extensions of (S, R).*

Proof. The CF2 extensions of (S, R) can be obtained as ordinary Dung extensions of (S, R_{CF2}). The latter has equational semantics involving elements of S only as variables. So this is the system of equations on S all of whose solutions give the CF2 extensions for (S, R). □

Also using a general theorem which says that we can eliminate higher level attacks, we can get another theorem:

Theorem 4.10. *Let (S, R) be a network. Then we can effectively construct a network (S_1, R_1), such that $S \subseteq S_1$, $R \subseteq R_1$ and the following holds:*

1. *Any CF2 extension E of (S, R) can be uniquely extended to a traditional Dung extension E_1 of (S_1, R_1).*

2. *For any traditional extension E_1 of (S_1, R_1) the set $E = S \cap E_1$ if non-empty is a CF2 extension of (S, R).*

Proof. Follows from reductions in [3]. □

Example 4.11. *To see how the equational approach works, let us compare the equations for 6 cycle with its symmetric closure (Figure 38).*

The equations for 6-cycle are:

1. $a_1 = 1 - a_6$

2. $a_2 = 1 - a_1$

3. $a_3 = 1 - a_2$

167

4. $a_4 = 1 - a_3$

5. $a_5 = 1 - a_4$

6. $a_6 = 1 - a_5.$

There are three solutions:

1. $a_1 = a_2 = a_3 = a_4 = a_5 = a_6 = \frac{1}{2}$ *(this is all undecided solution.)*

2. $a_1 = a_3 = a_5 = 1$
   $a_2 = a_4 = a_6 = 0$

3. $a_1 = a_3 = a_5 = 0$
   $a_2 = a_4 = a_6 = 1$

The symmetric closure of Figure 38 has the following equations:

1. $a_1 = 1 - \max(a_6, a_2)$

2. $a_2 = 1 - \max(a_1, a_3)$

3. $a_3 = 1 - \max(a_2, a_4)$

4. $a_4 = 1 - \max(a_3, a_5)$

5. $a_5 = 1 - \max(a_4, a_6)$

6. $a_6 = 1 - \max(a_5, a_1)$

The solutions (2) and (3) are also available here.

Example 4.12. *To conclude this section let us see how the network of Figure 19 will be adjusted by our method. Compare with Figure 23, which adds more nodes to the figure. The adjustment method does not add more nodes, but adjusts the attack relation R. Consider Figure 45.*

The adjustment is to make attacks in each of the loops $\{a, b, c\}$ and $\{x, y, z\}$ symmetrical. This took care of the CF2 maximal conflict free subsets of the loops. However, when either a or b are "in", d is also "in" and so x in the second loop is "out". The added symmetric attack $z \to y$ is redundant and wrong and should be disconnected. This is done by the higher level attacks $a \twoheadrightarrow (z \to y)$ and $b \twoheadrightarrow (z \to y)$.

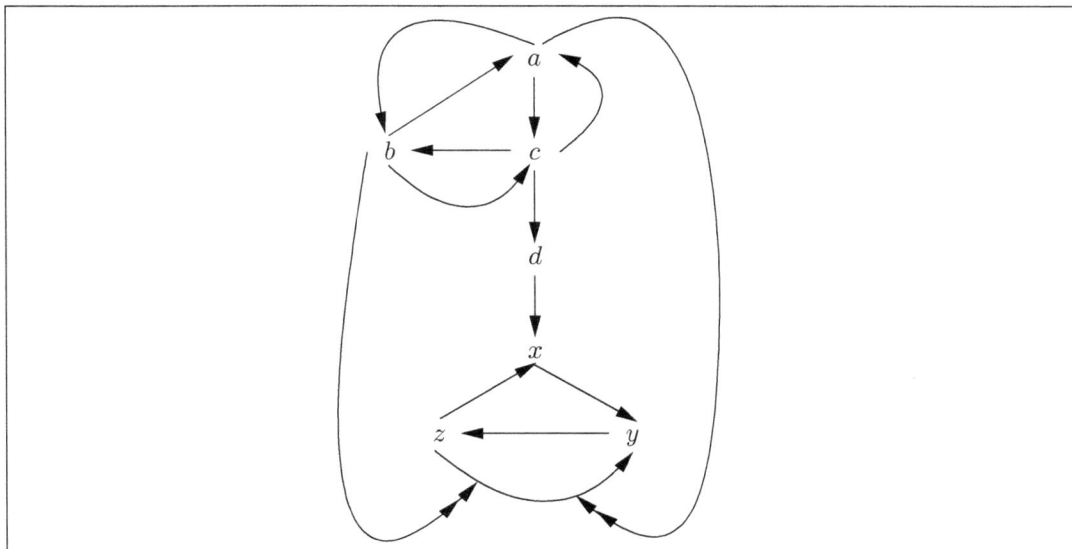

Figure 45:

5   Conclusion

We have seen various options for handling a family of argumentation networks arising from interactions between agents. We gave many detailed examples illustrating our options in calculating extensions for the overall system. In this concluding section we would like to discuss more the applicability of our methods and possibilities for further research.

Recall our motivating scenario of Section 1. We have two agents x and y. x is a religious fundamentalist, coming with an argumentation network (S_1, R_1) and with a complete extension E_1. y is a social liberal coming with network (S_2, R_2) and with extension E_2. They have to work together and so consider (S, R) with $S = S_1 \cup S_2$ and $R = R_1 \cup R_2$. The question is what extension $E \subseteq S$ can they agree on?

Our discussion in Section 1 related only to a starting strategy. We said that x cannot dismiss y and y cannot dismiss x, and each had better find another extension in the other network which is more favourable to him. So, for example, x will try to find a complete extension $E_{2,x}$ for (S_2, R_2) which minimally attacks the elements of his own E_1. Similarly y will find another extension $E_{1,y} \subseteq S_1$ which minimally attacks y's own E_2. This is a good start, where each is trying to work from within the point of view of the other. The question is how do we continue? We need another tool. The tool we want to use here is the Gabbay–Rodrigues iteration formula of [26]. This formula starts with an (S, R) with some initial values in the unit interval

169

[0, 1], and iterates the values until an equilibrium is reached. This equilibrium is a maximal admissible extension (not necessarily a complete extension) which respects the "in"=1, "out"=0 of the initial values. So what x and y can do is the following. Each puts initial values on the nodes that concern them and which they care about. These values can be in [0, 1] indicating how strongly they feel about the respective arguments. We average the values they put on each node to get the final initial values for the iteration algorithm and apply the Gabbay–Rodrigues iteration formula and get a compromise.

The adjustment method was used to get a representation for CF2 semantics. Its potential use should be further investigated. Given a network (S, R) and a preferred extension E, can we change R to R' so that (S, R') has the only extension E? We need to investigate how to change R to target only certain extensions. I believe Dvorak and Woltran [18] examined similar ideas with the use of extra points. We want to keep the arguments but only change the attack relation.

References

[1] P. Baroni, M. Giacomin, and G. Guida. SCC-recursiveness: a general schema for argumentation semantics. *Artificial Intelligence*, 168 (1-2):162–210, 2005.

[2] P. Baroni and M. Giacomin. Solving semantic problems with odd-length cycles in argumentation. In *Proceedings of the 7th European Conference on Symbolic and Quantitative Approaches to Reasoning with Uncertainty (ECSQARU 2003)*, pp. 440–451. LNAI 2711, Springer-Verlag, Aalborg, Denmark, 2003

[3] D. M. Gabbay. Fibring argumentation frames. *Studia Logica*, 93(2-3):231-295, 2009.

[4] Prabhaker Mateti, Narsingh Deo, On Algorithms for Enumerating All Circuits of a Graph *Siam Journal on Computing - SIAMCOMP*, vol. 5, no. 1, pp. 90-99, 1976 DOI: 10.1137/0205007

[5] Wolfgang Dvorak and Sara Alice Gaggl. Computation aspects of cf2 and stage 2 argumentation semantics. In *Proceedings of COMMA 2012*, pp. 273–284. IOS Press, 2012

[6] D. M. Gabbay. An equational approach to argumentation networks. *Argument and Computation*, special issue on the equational approach to argumentation, 3(2-3), 87–142, 2012.

[7] D. M. Gabbay. Semantics for higher level attacks in extended argumentation frames. Part 1: Overview. *Studia Logica*, 93(2–3), 355–379, 2009.

[8] P. Baroni, F. Cerutti, M. Giacomin and G. Guida. Encompassing attacks to attacks in abstract argumenation frameworks. In *Proceedings of the 10th European Conference on Symbolic and Quantitative Approaches to Reasoning with Uncertainty*, pp. 83–94, 2009.

[9] P. Baroni, F. Cerutti, M. Giacomin and G. Guida. AFRA: Argumentation framework with reursive attacks. *International Journal of Approximate Reasoning*, 42(1): 19–37, 2001.

[10] D. M. Gabbay and N. Olivetti. *Goal Directed Algorithmic Proof Theory* (Monograph), Springer, 2000, 266pp.

[11] M. Gelfond. Answer sets. In *Handbook of Knowledge Representation*. F. van Hermelen, V. Lifschitz, and B. Porter, eds., pp. 285–316. Elsevier, 2008.

[12] S. A. Gaggl and S. Woltran. cf2 semantics revisited. In *COMMA 2010*, volume 216, P. Baroni, F. Cerutti M. Giacomin and G. R. Simari, eds., pp. 243–254. IOS Press, 2010.

[13] S. A. Gaggl and S. Woltran. The cf2 argumentation semantics revisited. To appear in *Journal of Logic and Computation*, 2013.

[14] D. M. Gabbay. The equational approach to CF2 semantics. Short version. In *Proceedings COMMA 2012, Computational Models of Argument*, B. Verheij, S. Szeider and S. Woltran, eds, pp. 141–153. IOS Press, 2012.

[15] S. H. Nielsen and S. Parsons. Argumentation. A generalization of Dung's Abstract Framework for Argumentation: Arguing with Sets of Attacking Arguments. In *Multi-Agent Systems (ArgMAS)*, Nicolas Maudet, Simon Parsons, and Iyad Rahwan, eds. Future University, Springer 2006.

[16] P. Baroni, M. Caminada, and M. Giacomin. An Introduction to argumentatin semantics. In KnowledgeEng. Review 2011 01/2011, 26: 365–410.

[17] P. M. Dung. On the acceptability of arguments and its fundamental role in nonmonotonic reasoning, logic programming and n-person games. *Artificial Intelligence*, 77: 321-357 (1995).

[18] W. Dvořák and S. Woltran. On the intertranslatability of argumentation semantics, *J. Artif. Int. Res.*, 41(2), 445–475, 2011.

[19] W. Dvorák and C. Spanring. Comparing the Expressiveness of Argumentation Semantics. In *Fourth International Conference on Computational Models of Argument COMMA 2012*, pp. 261–272. IOS Press, 2012.

[20] T. J. M. Bench-Capon, S. Doutre and P. E. Dunne. Value-based argumentation frameworks. In *Artificial Intelligence*, pp. 444–453, 2002.

[21] D. M. Gabbay. The handling of loops in argumentation networks. To appear in *J. Logic and Computation*, special issue on loops in argumentation.

[22] R. Baumann and G. Brewka. Expanding argumentation frameworks: Enforcing and monotonicity results. In Baroni, P., Cerutti, F., Giacomin, M., and Simari, G. R. (Eds.), *Proceedings of the 3rd Conference on Computational Models of Argument (COMMA 2010)*, Vol. 216 of Frontiers in Artificial Intelligence and Applications, pp. 75–86. IOS Press, 2012.

[23] D. Gabbay and J. Woods. The laws of evidence and labelled deduction. Published in *Phi-news*, pp. 5–46, October 2003, http://phinews.ruc.dk/phinews4.pdf. Updated version in D.M. Gabbay P. Canivez, R. Rahman, and A. Thiercelin, eds., *Approaches to Legal Rationality, Logic, Epistemology,and the Unity of Science 20*, 1st Edition., Springer 2010,, pp 295-331 DOI 10.1007/978-90-481-9588-6_ 15.

[24] Soren Holbech Nielsen and Simon Parsons. *A Generalization of Dung's Abstract Framework for Argumentation: Arguing with Sets of Attacking Arguments in Argumentation in*

Multi-Agent Systems, Lecture Notes in Computer Science Volume 4766, 2007, pp 54-73.

[25] D. Gabbay. *Meta-Logical Investigations in Argumentation Networks*. Research Monograph, College Publications 2013, 770 pp.

[26] D. Gabbay and O. Rodrigues. Equilibrium States on Numerical Argumentation Networks, to appear.

[27] D. Gabbay and O. Rodrigues. A Numerical Approach to the Merging of Argumentation Networks Computational Logic in Multi-Agent Systems: 13th International Workshop CLIMA XIII Proceedings. Heidelberg: Springer, Vol. 7486, p. 195-212 (Lecture Notes in Computer Science). 2012.

[28] R. Baumann. Splitting an argumentation framework. In *Logic Programming and Nonmonotonic Reasoning*, J. P. Delgrande and W. Faber, eds., pp. 40–53. Vol 6645 of Lecture Notes in Computer Science, Springer, 2011.

[29] R. Baumann. *Metalogical Contributions to the Nonmonotonic Theory of Abstract Argumentation*. Dissertation, 2014.

Appendix

A   Higher order attacks — level 1 and joint attacks

This is a short note about how to handle level 1 attacks on attacks. See [7, 8] and [3, 25, 15].

Definition A.1. *An argumentation network with higher level 1 attacks has the form (S, R_1, R_2) where $R_1 \subseteq S \times S$ is attack relation and $R_2 \subseteq S \times R_1$ is the level 1 attacks on attacks. Figure 46 is an example of such a network.*

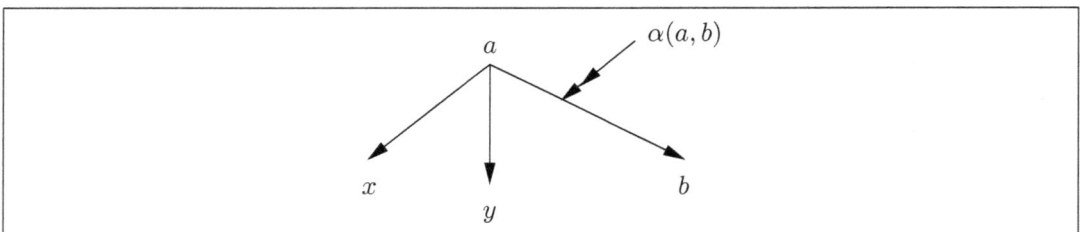

Figure 46:

We have

$$S = \{x, y, b, \alpha(a, b)\}$$
$$R_1 = \{(a, x), (a, y), (a, b)\}$$
$$R_2 = \{(\alpha(a, b), (a, b))\}.$$

The idea of providing semantics to such networks is to add new points in a conservative way and eliminate the attacks on attacks. Figure 47 shows how it is done for Figure 46.

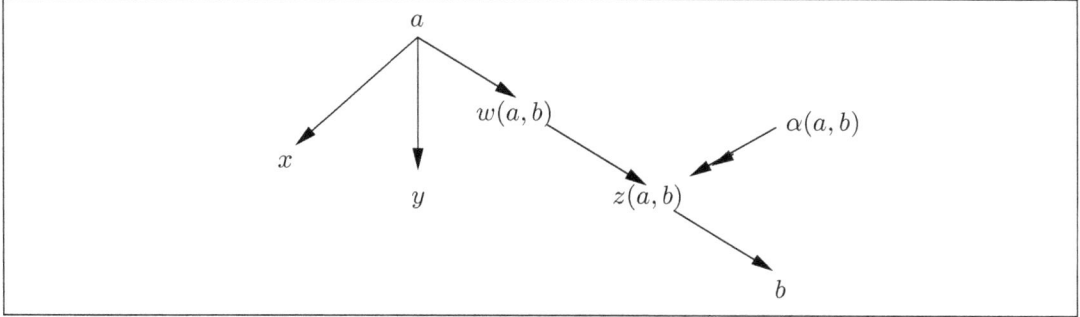

Figure 47:

What we do is to add new points $w(a,b)$ and $z(a,b)$ for any pair (a,b) such that a attacks b. To attack the attack $a \to b$ we can attack $z(a,b)$. It has the same effect.

Definition A.2. Let $\mathcal{A} = (S, R_1, R_2)$ be a network with level 1 higher attacks. We define $\mathcal{A}^*$, the flat companion of $\mathcal{A}$, as follows:

Let $\mathcal{A}^* = (S^*, R^*)$, where

$$
\begin{aligned}
S^* &= S \cup \{w(a,b), z(a,b)| \text{ for } (a,b) \text{ such that for some } c \in S, (c,(a,b)) \in R_2\} \\
R^* &= R_1 - \{(a,b)| \text{ for some } c \in S, (c,(a,b)) \in R_1\} \cup \{(c,z(a,b)), (a,w(a,b)), \\
&\quad (w(a,b), z(a,b)), (z(a,b), b)|(c,(a,b)) \in R_2\}
\end{aligned}
$$

Definition A.3. Let $\mathcal{A} = (S, R_1, R_2)$ be a level 1 higher network and let $\mathcal{A}^* = (S^*, R^*)$ be its flat companion. Then we say that a subset $E \subseteq S$ is a level 1 complete extension of $\mathcal{A}$ iff for some traditional complete extension of E^* of $\mathcal{A}^*$ we have $E = E^* \cap S$.

Definition A.4. An argumentation network with joint attacks has the form (S, R_3) where $R_3 \subseteq 2^S \times S$ where 2^S is the family of all subsets of S. When $S_0 \subseteq S$ attacks $s \in S$ (i.e. $(S_0, s) \in R_3$, we expect s to be "out" exactly when all elements $x \in S_0$ are "in".). See Figure 48.

Remark A.5. Joint attacks can be implemented using additional points as follows:

Given $(S_0, s) \in R_3$ add the new points $y_i = (S_0, x_i)$, for $x_i \in S_0$ and the point $\gamma(S_0, s)$ and the set of attack relations $R(S_0, s)$, where

$$
R(S_0, s) = \{(x_i, y_i), (y_i, \gamma(S_0, s)), (\gamma(S_0, s), s)|x_i \in S_0\}.
$$

173

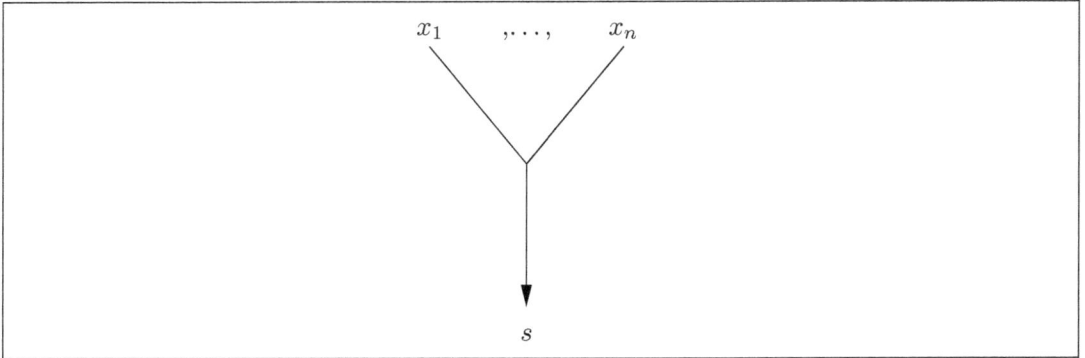

Figure 48:

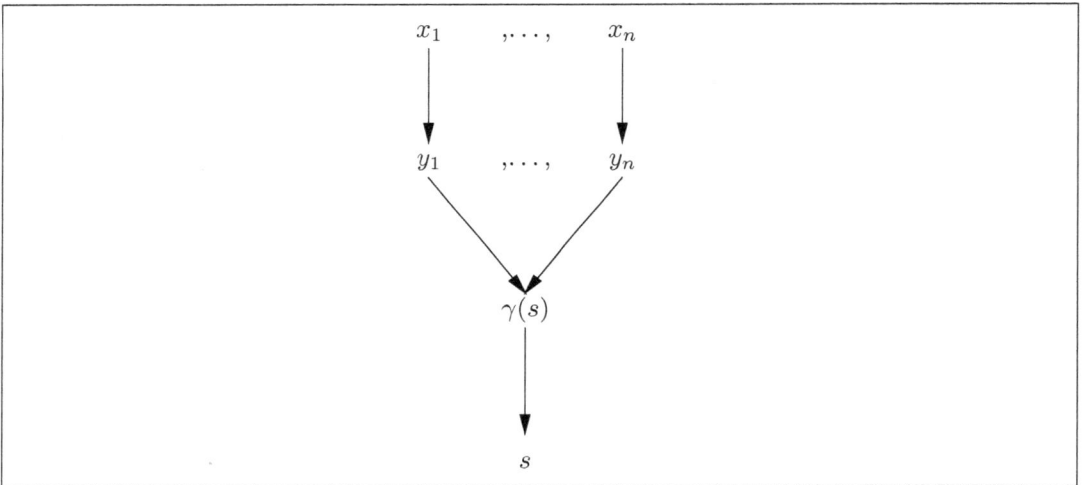

Figure 49:

We get Figure 49.

Then we have that s is "out" in Figure 49 iff all x_i are "in".

Therefore the constellation of Figure 48 is equivalent to the constellation of Figure 49.

This shows how joint attacks in (S, R_3) can be implemented in an extended network (S^*, R_3^*) with additional points of the above form and new attacks of the above form. That is, we define

$$S^* = S \cup \{y_i(S_0, x), \gamma(S_0, s) | x \in S_0 \ and \ (S_0, s) \in R_3\}$$

$$R_3^* = \bigcup_{(S_0, s) \in R_3} R(S_0, s).$$

174

Definition A.6. *Let* (S, R_3) *be a joint attacks network. Consider* (S^*, R_3^*) *as defined in Remark A.5. Then we define the complete semantics for* (S, R_3) *as all subsets of* S *of the form* $E^* \cap S$, *where* E^* *is a complete extension of* (S^*, R^*).

Remark A.7. *The level 1 higher level attacks and the joint attack concepts can be combined and we can define joint higher level attacks on nodes and arcs. We can look at networks of the form* (S, R_3, R_4) *where* $R_3 \subseteq 2^S \times S$ *and* $R_4 \subseteq 2^S \times R_3$. *An example of such a network is given in Figure 50.*

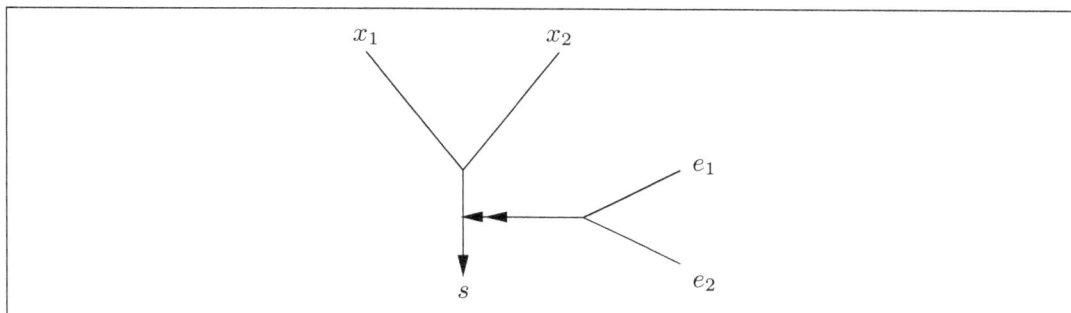

Figure 50:

This figure can be transformed to the following Figure 51 by first transforming the "joint" part and then transforming the higher level part.

If (S^*, R^*) *is the transformed network, we can again define* $E \subseteq S$ *is a complete extension of* (S, R_3, R_4) *iff for some complete extension* E^* *of* (S^*, R^*) *we have* $E = E^* \cap S$.

175

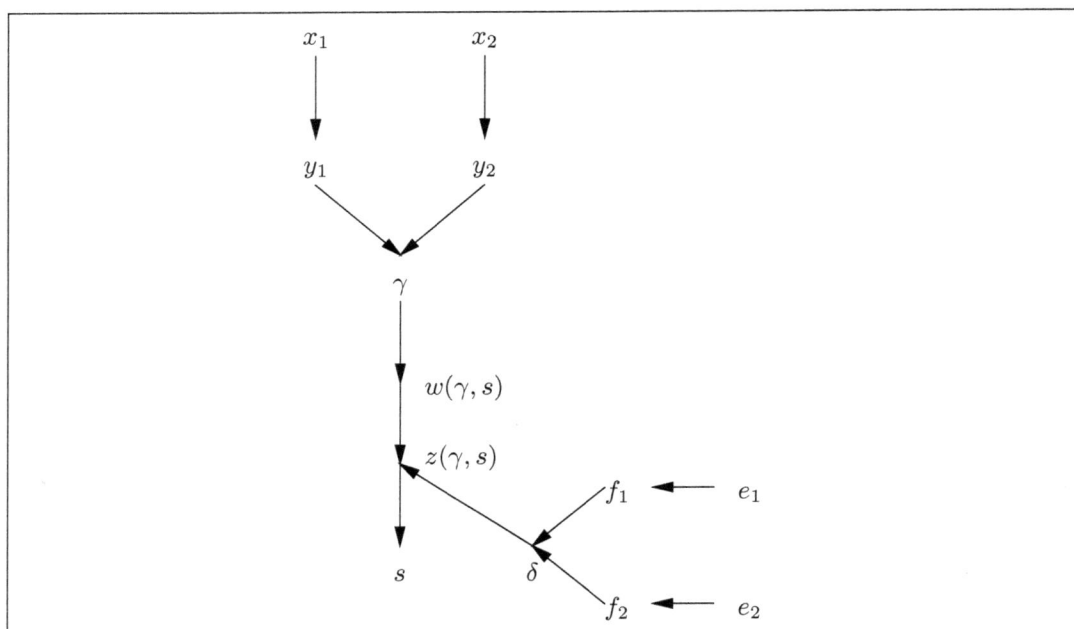

Figure 51:

Received January 2014

How Robust Can Inconsistency Get?

John Woods

University of British Columbia, Vancouver, Canada
john.woods@ubc.ca

"*Inconsistency robustness* is information system performance in the face of continually pervasive inconsistencies – a shift from the previously dominant paradigms of *inconsistency denial* and *inconsistency elimination* attempting to sweep them under the rug .... Inconsistency robustness is both an observed dominant phenomenon and a desired feature: It is an observed phenomenon because large information systems are required to operate in an environment of pervasive inconsistency. ...It is a desired feature because we need to improve the performance of large information systems."

Carl Hewitt

"We are all agreed that your theory is crazy. The question that divides us is whether it is crazy enough to have a chance of being correct." Neils Bohr to Wolfgang Pauli

Abstract

In 2011 Carl Hewitt introduced the concept of inconsistency-robustness (IR). IR is a property of information systems when the following conditions are met. The system is inconsistent, large, practical, inferentially self-aware, and the inconsistency is impossible to expel. Its inconsistency is also empirically discernible and a drag on the system's performance. Yet distinctively it is also the motivation occasion for performance improvement, albeit with no possibility of inconsistency-free successors. In this attenuated sense, inconsistency is also a desirable feature of IR systems. The technical details of IR are laid out in Hewitt's logic in progress IRL, itself an extension of his Direct Logic. It is possible that the idea of inconsistency-robustness is exhausted by the formal provisions of IRL. But I want to consider here a more latitudinarian possibility, according to which the idea of inconsistency-robustness admits of intellectually fruitful application well beyond the arcana of IRL. As we proceed, there will be occasion to close in on an interesting idea. It is that inconsistency has a more robust presence in human cognition than in Hewitt's large systems. More particularly, it is a presence unattended by cognitive dissonance. Concerning which, we conjecture that there is a plenitude of propositions unambiguously both true and false which cause no affront to the law of contradiction. Rightly enough, some will see this as a dialethic conclusion. But it is not to be found in the present-day logics of dialethic logic.

1   Some Technical Considerations

1.1   Inconsistency robustness logic

In this section and the two to follow I'll try to develop some sense of where Hewitt's idea of robust-inconsistency stands in the contemporary literatures of inconsistency-management. The idea of inconsistency-robustness takes fuller flight at the beginning of section 4.

Inconsistency Robust Logic (IRL) is an adaptation of Carl Hewitt's Direct Logic (DL).[1] DL is a logic that enables computers to carry out all inferences – including inferences about their own inferential processes – without the necessity of human intervention.[2] IRL is a logic built to accommodate the inconsistency-robustness claims as set out in the epigraph just above. For my purposes here it will suffice to confine myself to some remarks about IRL's DL component. In the case of inconsistency-pervasive[3] practical[4] theories, DL banishes proof by contradiction, as well as excluded middle and naïve ∨-introduction.[5] The Hewitt quotation makes two related assertions. One I'll call "the observability thesis" and the other "the desirability thesis." The observability thesis says that the pervasiveness and persistence of inconsistency are empirically discernible in the performance behaviour of large practical systems. The desirability thesis asserts that this inconsistency is a necessary and advantageous feature, made so by the fact that *"[c]urrent* systems

---

[1]For IRL see Carl Hewitt, "Inconsistency robustness in foundations: Mathematics proves its own consistency and other matters", a paper to be delivered at IR14, a conference on inconsistency robustness at Stanford University, summer 2014, date TBA. The epigraph is from an earlier draft of "Inconsistency robustness in foundations ..." and occurs almost word for word in Hewitt's "Inconsistency robustness for logic programs", also scheduled for IR14. Emphases in the text. For DL see Carl Hewitt, "Formalizing common sense for inconsistency-robust information integration using Direct Logic^{TM}", available at http://arxiv.org.abs/0812.4852.

[2]For example, without the further assistance of a software engineer.

[3]That is, the system's inconsistency is not just one-off. Rather it has a more liberal distribution, but by no means necessarily a majority or even dominant presence.

[4]A practical software system typically has tens of millions of lines of code. Examples of practical systems are climate models and diagnostic and treatment systems in medicine.

[5]However, it accepts double negation and the other Boolean equivalences.

do not perform adequately in the face of inconsistency, and must be improved."[6] Hewitt asserts that large inconsistent systems work better with the inconsistencies left in − that is to say, robustly managed − than they do with the inconsistencies swept under the rug – or denied entry in the first place.[7] When these two conditions are met, there is an allowable extension of the term in which it is the information system itself that is inconsistency-robust.[8]

Of particular interest here is Hewitt's proof of the self-proved consistency of mathematics, informally set out as follows.

1. Suppose for contradiction that mathematics is inconsistent.

2. So there is some proposition Φ of mathematics such that $\vdash \Phi$ and $\vdash \sim\Phi$.

3. From this we have an immediate contradiction: $\ulcorner\Phi \ \wedge \sim\Phi\urcorner$ is a theorem of mathematics.

4. So by *reductio* mathematics is consistent.[9]

If the proof stands up, it would now seem that we have an easy way to prove not the relative consistency of mathematics but its absolute consistency as well:

a. Either mathematics is consistent or inconsistent.

b. If consistent, well and good.

c. If inconsistent, then consistent.

d. So mathematics is consistent.

---

[6]Hence its persistence. Hewitt to Woods, personal communication, 31 August 2013.

[7]Indeed "[p]ervasive inconsistency cannot be removed from large software systems". Emphasis added. Hewitt to Woods, 31 August 2013.

[8]The idea that inconsistency can be an attraction for a system does not originate here. An earlier favourable mention can be found in, e.g., Dov M. Gabbay and Anthony Hunter, "Making inconsistency respectable part 1", in Ph. Jourand and T. Kelemen, editors, *Fundamentals of Artificial Intelligence Research,* volume 535, pages 19-32, Berlin: Springer-Verlag, 1991, and Dov M. Gabbay and Anthony Hunter, "Making inconsistency respectable part 2: Meta-level handling of inconsistency", in *Lecture Notes on Computer Science 747*, pages 129-136, Berlin: Springer-Verlag, 1992 pages 306-327. See also Peter Schotch and Raymond Jennings, "On detonating", in Graham Priest, Richard Routley and Jean Norman, editors, *Paraconsistent Logic: Essays on the Inconsistent*, pages 306-327. Munich: Philosophia, 1989.

[9]Hewitt, "Mathematics self-proves its own consistency (*contra* Gödel *et al.*)". As regards line (3), Hewitt writes: "Consequently both Φ and $\neg\Phi$ are theorems that can be used in the proof to produce an immediate contradiction." (p. 1)

The trouble is that line (a) calls upon excluded middle, which Hewitt excludes from DL and IRL.

There is an even more threatening generalization of it to consider. Let Θ be an arbitrarily selected theory and let $\vdash_\Theta$ denote the relation $\ulcorner\Phi$ holds in $\Theta\urcorner$. Then

1′. Suppose for contradiction that Θ is inconsistent.

2′. So there is some proposition Φ such that $\vdash_\Theta \Phi$ and $\vdash_\Theta \sim\Phi$.

3. From this we have an immediate contradiction: $\vdash_\Theta (\Phi \wedge \sim \Phi)$.

4. So by *reductio* Θ is consistent.

But again,

a′. Either Θ is consistent or inconsistent.

b′. If consistent, well and good.

c′. If inconsistent, then consistent.

d′. So Θ is consistent.

Since Θ is arbitrary, we have it then that

e′. Every theory is consistent.

This is rather striking. For anyone at ease with excluded middle, Hewitt's generalized proof does for consistency the sort of thing the Curry-proof did for truth.[10] That is, it proves that there is massively too much of it. This we might call "the paradox of consistency."

I don't want to overpress the Curry-proof resemblance. It hinges on my version of Hewitt's own informal presentation of his proof. It does not hinge on the formal presentation of Hewitt's proof as set out in DL. Hewitt's informal proof is set out in non-technical English. My generalization of it is too. My Hewitt's proof invokes rules and assumptions that fail in Hewitt's own DL and IRL. Given their respective purposes, in DL and IRL, Hewitt might well be motivated to make them fail there. But this is a long way from meaning that they should also fail for non-technical English. This qualifies the resemblance of my proof of omni-consistency to Curry's proof of omni-truth. But it does not obliterate it. When all is said and done, Curry's

---

[10] Haskell B. Curry, "The inconsistency of certain formal systems", *Journal of Symbolic Logic, 7* (1942), 115-117.

is a proof about the English predicate "true", and mine is a proof about the English predicate "consistent". But if the crudity of saying so might be forgiven, the *guts* of my generalization of Hewitt's informal proof are the guts of that proof too.

A second difficulty with the proof is that even if valid it does not appear to establish that it is mathematics itself that proves its own self-consistency. If so, it is hard to see how, as Hewitt himself avers,[11] it engages with, much less contradicts, the Gödel second incompleteness theorem. Of course the present formulation of the proof is a loosely informal one. The actual proof lies deep in the coils of DL.[12] The language of DL is purpose-built to pronounce upon its own doings, in ways that enable DL to prove theorems about itself. Assume now an extension of the language of DL – call it FA_{DL} – which gives formal expression to the axioms of first-order arithmetic in a way that provides that if $\vdash \Phi$ and $\vdash \psi$, then $\vdash (\Phi \land \psi)$. Then as presented there, Hewitt's proof must look something like this: If FA_{DL} is inconsistent, ⌐FA_{DL} is consistent⌐ is provable in FA_{DL}. Supposing that all goes well, it still isn't clear that Hewitt's formal proof conflicts with Gödel's. Gödel established that if FA is consistent, it is not provably so in FA. Of course, if excluded middle were allowed[13] the desired incompatibility would be solidly established, but only at the expense of a Curry-like result for consistency.

The Curry-like result also carries consequences for Hewitt's more immediate concerns. If proof by contradiction were an allowable rule for arbitrary Θ[14], there would be no such thing as inconsistency *robustness*. If anything's robust here, it is *consistency* and in so saying conferences on inconsistency are now denied a coherent motivation. Hewitt is right to say that if inconsistency robustness is to be saved, it is essential that IRL suppress proof by contradiction. But robustness aside, if we merely wanted to give the plain old predicate "is inconsistent" a non-null extension, we would also have to disable proof by contradiction.[15] Thus the reach of Hewitt's

---

[11]Kurt Gödel, "On formally undecidable propositions of *Principia Mathematica* and related systemsI", in Solomon Feferman, John W. Dawson, Jr., Stephen C. Kleene, Gregory H. Moore, Robert M. Solovay and Jean van Heijenoort, editors, *Kurt Gödel's Collected Works*, volume 1: *Publications 1929-1936,* pages 145-195, New York: Oxford University Press, 1986.

[12]"Mathematics self-proves . . . ", p. 1. Adjunction, also known as $\land$-introduction.

[13]Of course, it does hold in first-order Peano.

[14]Not forgetting that *in DL* proof by contradiction fails for practical theories that are pervasively-inconsistent.

[15]Alternatively, we could give up on adjunction. There are lots of logics – in which the adjunction rule fails, beginning with the discussive logic of Stanisław Jaśkowski. Indeed non-adjunctive logics are a standard way to make logic paraconsistent. Non-adjunctive logics are also called preservationist. Some people think that "non-adjunctive" is a slighting term for preservationism. But let that pass for now. An excellent introduction is Peter Schotch, Bryson Brown and Raymond Jennings, editors, *On Preserving : Essays on Preservationism and Paraconsistent Logic, Toronto*: University of Toronto Press, 2009. See especially chapter 6, "Preserving what?" by Gillman Payette

proof considerably outruns his preoccupation with inconsistency robustness.

We might think that a proof causing this much trouble is too delicious to be true. That certainly is Hewitt's own view of the matter. Hewitt does *not* think that his proof causes trouble on any scale that would remotely approach the trouble caused by a Curry-like result for consistency. The reason for this is that he thinks that his proof doesn't generalize in the way that I've suggested. What he does think is that the proof succeeds in showing that "it is *only* in practical pervasively inconsistent theories that" proofs by contradiction fail.[16] However, as Hewitt insists, it is the practical inconsistency-pervasive theories that are important for large systems.

My generalized version of Hewitt's informal proof gives a result of sufficient excessiveness to justify the efforts to disarm it. One possibility is the refusal of proof by contradiction rule, notwithstanding its dominant and necessary employment in the existing consistency proofs in standard mathematics. Another is the repudiation of adjunction. Since I myself am not much inclined to give up on adjunction, it would appear that proof by contradiction will have to go, in spite of Hewitt's insistence that the proof by contradiction rule fails only for robustly inconsistent systems. So perhaps the present option requires further consideration.

Whether or not Hewitt is right in saying that *my* generalization of Hewitt's informal proof is not a generalization of Hewitt's informal proof, I daresay that some readers will think that my proof is a validly serious discouragement of proof by contradiction beyond the confines of inconsistency robust theories. It certainly matters whether my Hewitt's proof stands or falls. If it stands, inconsistency robustness doesn't bear thinking bout. If it doesn't fall, we'll have to get clearer about how Hewitt's Hewitt's proof attains its exclusive attachment to IR contexts. That is not evident in Hewitt's informal exposition of his proof. To get to the bottom of things, we would have to study the formal intricacies of IRL. But that is not my mission here.[17]

Still, I don't mean to make light of this matter. If perchance my generalized version of the proof were right, we would have to proceed with care. We would have to take pains to honour a distinction between a proof by contradiction and a *reductio ad absurdum* proof. Here is why. The case against proof by contradiction afforded

---

and Peter Schotch (pages 85-104). Cf. Graham Priest, "Paraconsistency and dialetheism", in Dov M. Gabbay and John Woods, editors, *The Many Valued and Nonmonotonic Turn in Logic*, volume 8 of their *Handbook of the History of Logic*, pages 129-204, Amsterdam: North-Holland, 2007. See also Bryson Brown, "Preservationism: A short history" at pages 95-128 of the same volume. I will say later why I think the adjunction rule isn't a deal-breaker here.

[16] Carl Hewitt, personal correspondence, 24 August 2013; emphasis added.

[17] Further reservations about Hewitt's proof are set out in an appendix to my "Inconsistency: Its present impacts and future prospects", scheduled for the IR14 conference at Stanford in the summer of this year.

by *my* generalization of Hewitt's proof is itself advanced as a *reductio* argument, summarizable as follows:

1. Proof by contradiction is a valid rule.

2. Provided proof by contradiction is a valid procedure, it can be demonstrated that everything whatever is consistent.

3. But this seems absurd.

4. So proof by contradiction is not a valid rule.[18]

All this is very interesting and important. But since my concern here – like Hewitt's own – is with inconsistency-*robustness*, I'll turn to that now, leaving the intricacies of *reductio* reasoning and related matters for another time.

1.2   Is there a DL tie to paraconsistency?

In Hewitt's approach, IR systems are those in which, although inconsistencies cannot be expunged, the systems remain operational. It is hard to see that this could be so unless a theorem of classical note were disarmed. A system is negation-inconsistent just in case for some sentence Φ, both $\vdash \Phi$ and $\vdash \sim\Phi$. It is generally agreed that the defining feature of paraconsistent systems is their rejection of the *ex falso quodlibet* theorem, by which negation-inconsistency and *absolute* consistency are strictly equivalent properties; in particular, by which everything whatever can be proved from a negation-inconsistency. We see, then, that, *ex falso*-compliant systems *detonate* in the presence of any contradictory pair of theorems; which is the very thing that paraconsistentists want to avoid. This raises the question of how a paraconsistent system should manage its negation-inconsistencies. A dominant answer bids us to find (or define) one or more premiss-conclusion relations for which *ex falso* fails but which is *property-preserving* with respect to some desirable property or properties other than truth. In some approaches this property to be preserved is "bad, but not too bad". Its preservation requires that in its management of the logic things are not made worse. If *ex falso* held true, this would be a forlorn hope. Negation-inconsistency is bad, but not too bad if it is carefully treated. But in *ex falso* logics, it is handled in ways that make everything inconsistent. That's bad too, but shockingly worse.[19]

---

[18]There is a less venturesome version of this *reductio* got by adding to the end of line (1) the words "in classical mathematics", and to the end of line (4) "for proving the consistency of mathematical theories." Still, it's a bad result.

[19]Cf. Hippocrates: *"Primum non nocere"*. For more see Peter K. Schotch, "Paraconsistent logic: The view from the right", *Philosophy of Science Association*, 2 (1993), 421-429.

Paraconsistent logics divide fairly neatly into two camps. In the one, *ex falso* is blocked by imposition of relevance constraints on entailment and deducibility. In the other, *ex falso* is blocked in some other way, for example, by suppressing the rule of adjunction, without the need to impose a relevance condition. See, in this regard, Nicholas Rescher and Robert Brandom, *The Logic of Inconsistency*.[20] At the time of its publication, most of the paraconsistent work done by American logicians was of the first type. Indeed some of the leading relevant logicians were Rescher's colleagues at the University of Pittsburgh. But Rescher himself, and Brandom too, are in this second camp, in the slipstream of paraconsistent developments in Poland and Brazil. The paraconsistent terrain in Australia tended to subdivide into dialethic logic and relevant logic, with some concomitant overlap. The paraconsistent turn in Canada is substantially influenced by preservationist adaptations of Rescher and Brandom, got by weakening some of the latter's inferential constraints.[21] Also important are Belgium's contributions to adaptive logic. The Rescher and Brandom essay is a response to two different impulses. One is the impulse to free set theory from the Russell contradiction, and to do so with least possible disturbance of the theory's original intuitions. The other is to have a logic that is duly sensitive to the particularities of assertion. Accordingly we might propose that

> *Ex falso for* assertion: A relation that preserves theoremhood would seem to be more *ex falso*-friendly than a relation that preserves assertion.

Although these are clearly inequivalent objectives, they embedded a common strategy. It entailed, in the manner of Jáskowski, the suppression of the adjunction rule, the rule that permits unrestricted premiss-aggregation. This disables the classical equivalence of $\{A, B, C, \ldots\} \vdash X$ and $\neg A \wedge B \wedge C \wedge \ldots \neg \vdash X$.[22]

Towards the end of *The Logic of Inconsistency*, we find the idea of an "intelligible" formal system. Clearly a technical term, a formal system is intelligible just in case it meets the following three conditions. (1) Its theorems include all the logical truths. (2) No self-contradiction is a theorem. (3) Any classical consequence of a theorem is a theorem. However condition (3) doesn't apply to *sets* of theorems. An intelligible formal system not only isn't closed under conjunction, it may also prove contradictory pairs of theorems (but not their conjunctions).

---

[20] Oxford: Blackwell, 1980.

[21] For the tie between preservationist logics and Rescher's and Brandom's brand of non-adjunctive logic, see Bryson Brown, "Rational inconsistency and reasoning", *Informal Logic*, 14 (1992), 5-10. The prservationists' pivotal notion of forcing isn't, however, prefigured in the Rescher-Brandom approach.

[22] S. Jáskowski, "On the rules of supposition in formal logic", in Storrs McCall, editor, *Polish Logic: 1920-1939*, Oxford: Oxford University Press, 1967; first published in Polish in *Studia Logica*, 1934.

These days, there is a hefty multiplicity of paraconsistent logics, mainly inequivalent and often one another's rivals. Comparing DL with paraconsistent logic *as such* is too much to hope for. So far, there is only one of it, and too many of them. Still some things are clear. As we saw, DL proves the self-consistency of formal arithmetic. Again, the proof informally rendered is this.

1. Suppose for reductio that FA is inconsistent.

2. For some $\Phi, \vdash_{FA} \Phi$ and $\vdash_{FA} \sim \Phi$ from 1, by def. of inconsistency

3. $\vdash_{FA} (\Phi \wedge \sim \Phi)$ from 2, by adjunction

4. FA is consistent.   from 3, by proof by contradiction

One of Hewitt's tasks is to disable this proof. It is interesting to take note of the remedies he *doesn't* apply here. Unlike all non-adjunctive paraconsistent systems, Hewitt doesn't suppress adjunction. Unlike most breeds of paraconsistentism, he doesn't press the objection that if Φ and $\ulcorner \sim \Phi \urcorner$ were allowed to stand as theorems of FA, FA would be up to its eyeballs with endlessly more contradictory theorempairs. Nor, unlike relevant paraconsistents, is there any overt concern with *ex falso*. For Hewitt, the overriding fact is that FA proves an outright contradiction which is in immediate and direct contradiction of the law of non-contradiction LNC. This is a point of pivotal importance, dispossessing all other concerns of a place at the table. If FA proves Φ and $\ulcorner \sim \Phi \urcorner$, it violates LNC, and that fact alone is wholly decisive; it constitutes the entire case against it. Whereupon it is immediate that FA is not inconsistent.[23]

Later on I will have something to say about Aristotle's version of his own non-contradiction law. All we need say about it now is that there is no trace of Aristotle's interest in the equivalence or otherwise of contradictory pairs of propositions and

---

[23]With permission I quote a helpful anonymous referee: "When we apply RAA, there are three possibilities. 1. We already had a contradiction inconsistency (amongst the premises accepted up to the point of the sub-derivation). 2. Adding the hypothesis of the sub-derivation took us "over the edge". 3. Both. In the first case, the premises already in place are inconsistent and (classically) *ex contradictione quodlibet* gets us the 'rejection' of the hypothesis. In the second case, the premises already in play are inconsistent with the hypothesis, so the conclusion of its negation from the premises in place is correct. In the third case, we have an embarrassment of riches – the premises on the table are enough on their own to get a contradiction, and we add a further premise to the mix which is independently troubling. The result, in every case, is classically valid derivations – but the conclusions we might be tempted to draw from that derivation need careful handling, since the three cases differ sharply in terms of the logical situation they involve. We all know that hanging an innocent proposition for a contradiction we already had in our hip pocket can be a handy manoeuvre, but it doesn't show anything logically 'bad' about the victim!"

contradictory conjunctions of them. Such textual support as we have indicates that the closest Aristotle comes to the modern LNC is in a contradictory- pairs formulation of it.[24]

Hewitt's remedy is to retain the contradictory conjunction reading while concurrently disabling the proof by contradiction rule. It is a risky move, as no one knows better than he. The proof by contradiction rule is indispensable for mainstream mathematics. Hewitt will need to do two things at once. He will need to keep the proof rule in general circulation while clipping its wings here. The desired accommodation will be provided in IRL, which is a purpose-built adaptation of DL. IRL has a number of interesting properties. One is its suppression of self-reference. Another is the welcome it extends to pervasive and wholly entrenched inconsistencies in very large information systems. In the first instance, IRL takes a quite orthodox approach to inconsistency-producing self-reference. It banishes self-reference. In the second, it is in a league by itself.

1.3 Thought-to-be consistency

Hewitt is strongly of the view that the proof by contradiction rule is both allowed by and is indispensable for "thought-to-be consistent" mathematical theories. Suppose, however, a given thought-to-be-consistent theory in fact were inconsistent. Hewitt is adamant that "you can't prove everything." But this is precisely what you can do if *ex falso* is true. Everything could be proved if the system has $\vee$-elimination, $\wedge$-introduction, and double negation. DL disallows $\vee$-introduction but admits the others. Would that make DL a paraconsistent logic?

Perhaps not. Might we not get *ex falso* in other ways, and more directly, from the definition of $\models$: $\Phi \models \psi$ iff $\sim \diamond (\Phi \wedge \sim \psi)$. If the system is sound and complete, the same holds for $\vdash$: $\Phi \vdash \psi$ iff $\sim \diamond (\Phi \wedge \sim \psi)$. Putting $\ulcorner A \wedge \sim A \urcorner$ for Φ and $\sim B$ for ψ, we now have it that $A \wedge \sim A \models \sim B$, for any B. Since double negation holds in DL, this gives it that $A \wedge \sim A \models B$. Wouldn't this make DL a *non*-paraconsistent detonating logic? If so, wouldn't this also imperil Hewitt's thought-to-be-consistent rescue of proof by contradiction for all systems, except for those very particular ones in which it is excluded as a matter of policy?

A related issue concerns IR systems. IR systems are not only inconsistent but are discernibly so. They wear their inconsistencies on their sleeves. If, as suggested, DL allowed *ex falso*, here too IR systems would detonate, and everything would be provable there, never mind whether proof by contradiction is also allowed. Hewitt says that DL must allow proof by contradiction for all systems thought-to-be consistent. That way the business of mathematics can be transacted *comme d'habitude*.

---

[24]Perhaps also in some of the Fitch systems.

In fact, however, DL avoids these difficulties, on a technicality we might say. DL isn't a model theoretic logic and has no capacity (or wish) to express or regulate "$\diamond$" and "$\models$". Why would we say that this problem is avoided "only on a technicality"? I say it because the modal definition of $\models$ might be true and the derivation of *ex falso* might follow from it validly. If these things actually were so, their not being so in DL would cause them no discouragement, but wouldn't it be fairly called into question whether DL is the logic we should want for mathematics?

What does it take to fulfill the "thought-to-be" condition? If it means "is, for the sake of mathematics, taken to be" or "is, for the sake of mathematics, stipulated to be", then why wouldn't we apply this face-saving measure even to systems that are discernibly inconsistent? Hewitt's answer is that this not what the thought-to-consistent qualification comes to. "By a 'thought-to-be consistent mathematical theory', I mean one that is thought to be consistent by the overwhelming consensus of working professional mathematicians."[25] Examples include the second order axiomatizations of the integers and reals, Hilbert spaces, and group theory. So it is clear that the final occurrence of "thought" in the quoted sentence means "widely believed by the experts."

Hewitt thinks that when you discover that a thought-to-be consistent mathematical theory is actually inconsistent, tools must be downed until repairs are made. An inconsistent thought-to-be consistent theory is a wrecked theory and must stay wrecked until ways are found to make the inconsistency go away. Inconsistency-*deniability* is the required goal here. IR theories are a thing apart. They leave the thought-to-be consistent strategy entirely without motivation.

IR systems are differently purposed. Their job is not the paraconsistentist's job to keep its inconsistencies strictly in their place by giving them a neutered operational presence there. The job of an IR systems is to give its inconsistencies freer rein and more productive play. The question I raise here is whether Hewitt's understanding of robustness, is the way to give optimal expression to the interesting idea of inconsistency robustness should *ex falso* be true, never mind that it isn't so-acknowledged by DL itself. But I shan't press the matter here. As I said, what catches my present interest is the *very idea* of inconsistency robustness.

1.4   Three grades of IR involvement

I come now to the central question of this paper. Does the idea of inconsistency-robustness have legs to stand on its own in contexts other than the technical provisions of IRL? Mine will be an answer in the affirmative. Inconsistency-robustness is unarguably a property of absorbing interest. Its attractiveness makes one want to

---

[25] Hewitt to Woods, personal communication, 10 September 2013.

know more, especially about the robustness part. Robust inconsistency, as we saw, is a property of information systems definable over theories. At this level of mathematical abstraction there is no presumption that a system actually believes its data, indeed that systems are believers of any kind. Let us say then that the first grade of inconsistency-robustness is the inconsistency robustness of *presence*. The inconsistent data are present and settled into the system at hand. To achieve the second grade of robustness, believers must now enter the picture in a load-bearing way. Information systems would now be intelligent interactors capable of inconsistency-transparent belief and (doxastic) self-awareness. When such a system is large,[26] its inconsistency extensive and self-awarely believed, it will exhibit the inconsistency robustness of *belief*.

The third grade is achieved by admitting the rest of the world into the picture, by giving reality itself a load-bearing presence in the inconsistency-robustness story. The world achieves factual robustness when enough of the robust inconsistencies believed by its interactive agents are actually true. This is the inconsistency-robustness of *being*. In our ascent from robustness of presence, to doxastic variations of it, and thence to the ontic robustness of inconsistent facts, it is clearly the last step that is the philosophically most audacious. But before getting on with its further consideration, I'd like to register an observation about grade-two robustness. The robustness of believed inconsistencies is a considerable phenomenon in its own right, carrying with it real illumination about inconsistency-pervasiveness and genuine instruction about the virtue of keeping inconsistencies doxastically secure. Even if we strike out with grade three robustness, grade two success is no booby-prize.

An especially interesting possibility to consider is a working partnership among the three grades. Might it be the case that there actually are cases in which there are masses of inconsistent data, masses of them are believed true, and a goodly portion of those indeed are true? If so, let us say that any intelligent agent who is capable of the coherent management of such a partnership is a *three-grade* manager of robust inconsistencies. The obvious question is whether three-grade managers actually exist. In what follows I'm going to sketch a foundation for answering the question affirmatively. I will explore the idea that three-grade managers are a reality and that we humans are among their better practitioners.

It will strike some readers as foolhardy, or worse, even to consider a move from grade-two inconsistency robustness to grade three. Surely, they will say, it is one thing (and already a stretch) to acknowledge the robustness of inconsistent beliefs, but the actual inconsistency of the world is simply beyond consideration. Against

---

[26]Of course, there is large and then again there is large. The human animal may not be a Big Data manager, but given his nature and his circumstances he's up to his ears in data – and then some.

this I want to say that anyone who has the time of day for doxastic robustness is, just so, at least half the way to ontic robustness, and therewith to a favourable view of three-grade inconsistency management.

Key to it all is the *semantic purport* of ready belief, according to which a person who really *believes* that Φ really believes that Φ is *true*. Second grade inconsistency management is the management of pervasive inconsistencies in a system's belief-set. When a belief-set is one that is widely shared in a human population, then there are inconsistencies that virtually all its inhabitants believe; and if those inconsistent beliefs are pervasive, that alone is evidence of the adequacy of the belief-set's *management*. Take at random any member of this vast horde of inconsistency-believers. This is a person who believes it to be true that many inconsistent sentences are true. Suppose now that a metaphysically inclined interlocutor presents himself and says, "Perhaps we might now consider what makes it the case that what you believe to be true is indeed true". Suppose that this provokes the reply: "The very idea of it is absurd. You're asking me to take it seriously that contradictions might actually be true!" The reply to this can only be, "but isn't that what you already believe to be so?"

The moral of this exchange is that anyone at home with the notion of the inconsistency-robustness of belief is, by the semantic purport property, already committed to the inconsistency-robustness of the world. So it would be the most transparent kind of special pleading to acknowledge one's membership in the managerial class of grade-two inconsistency robustness but disdain the very possibility of a shift upwards to grade-three.

There are three claims about systems management currently under consideration. One is that the presence of inconsistency-robustness in a large system is of practical assistance in the system's management. The second says the same thing for robustly inconsistent systems of belief. The third extends this to robustly inconsistent systems of true belief. All three are theses about inconsistency-management. Each is an answer to the question of how inconsistencies are to be managed. The question posed by the third is, "How is a human agent to handle himself in a world that is actually inconsistent?" The answer, I suggest, is little more than this: The best way for a human agent to handle himself in an inconsistent world is to figure out how to handle himself as a *believer* in the world's inconsistency. In shorter words, the best way to be a good grade-three manager is by being a good grade-two manager.

My task is to determine whether the notion of three-grade management deserves a shot in the context of the management of large information systems possessed by humans. I will be the first to concede that if in the rich flow of the intelligent interactions of humans, we can't find plausible candidates for consideration, then that would be sufficient cause to vacate my question. So we must ask: "Where are the examples?" The place of *fiction* in human cognitive life offers some fertile ground.

2 Some Philosophical Considerations

2.1 Fiction

The logic of fiction has been around since the early 1970s. It has achieved some recognition in philosophical theories of reference. It achieves mainly passing mention in free logics and existence-neutral quantificational systems. More recently, the idea of fiction has enjoyed a small philosophical renaissance. It is the foundational strut in all the various factionalisms that we find in the philosophy of mathematics, science, metaphysics, ethics, and who knows what else? We have in our own present considerations occasion to invoke it further. Fictions play a large role here. They will be a primary datum for IR reflections on a scale that involves us all.

I begin with a word of defence of the importance and relevance of fiction, with Frege as my point of departure. Frege was one among a number of important people who took the position that fiction doesn't matter for anything that mattered philosophically. What mattered to Frege philosophically were the foundations of logic and mathematics. In his curt dismissal of fiction, Frege makes it clear that he thinks that a semantics for fiction doesn't matter for what actually does matter.[27] Indeed it wouldn't even matter whether fictional sentences had or lacked any semantic character at all.[28] It wouldn't matter therefore how fiction is treated (if dealt

---

[27] Gottlob Frege, "On *Sinn* and *Bedeutung*", trans. Max Black, in Michael Beaney, editor, *The Frege Reader*, pages 151-171, Oxford: Blackwell, 1997; first published in German in 1892. Frege held that since the sentences of fiction lack truth values, "we are interested only in the sense of the sentences and the images thereby involved", and that the question of truth is not involved in a work of art, but only in scientific investigation. (p. 163) In "On denoting", Russell has even less to say about fiction. He proposes that fictional names be replaced with descriptions in which full scope renders all fictional sentences false. ("On denoting", in Alastair Urquhart, editor, *The Collected Papers of Bertrand Russell*, volume 4: *The Foundations of Logic*, 1903-05, pages 415-427, London: Routledge, 1994; 425; first published in 1905 in *Mind*.) He is more expansive in *An Introduction to Mathematical Philosophy*, London: Allen and Unwin, pages 169-170, 1967; first published in 1919. But the discussion is thin and naïve, and betrays a lack of serious effort. Strawson's fleeting mention of fiction is even less serious (P.F. Strawson, "On referring", *Mind*, 59, 1950, 320-344). In the original version Strawson calls the fictional uses of referring expressions "spurious". In later versions he replaces "spurious" with "secondary", without a jot of explanation as to its meaning here.

[28] From a somewhat broader perspective, Frege's indifference to fiction is its incapacity to advance our knowledge, not only of logic and the foundations of mathematics but also of the subject matter of any kind of knowledge-seeking science. He anchors this objection to his insistence that the sentences of fiction lack truth values of any kind. See "on *Sinn* and Bedeutung", p. 159.

with at all), provided it didn't discomport itself with what matters to Frege. Part of what I want to do here is to show how premature and mistaken Frege's dismissal was[29], that fiction not only does matter for what matters to Frege, but it should also matter to those for whom inconsistency-robustness matters.

To get things started, I want to linger awhile with an interesting *empirical* fact. It is that, except for specialized tutelage, virtually everyone on earth who has read a novel or short story with understanding[30] believes without anxiety the first five lines of the argument that now follows. These lines are the set-up of what is frequently described as "the paradox of fiction".

1. There are sentences that refer to Sherlock Holmes and the events of his life. ("Sherlock Holmes lived at 221B Baker Street, in London".)

2. Some of these sentences are true. (See above)

3. Sherlock Holmes doesn't exist and no event in his life is an event that exists, then or now or ever.

4. It is not to Sherlock Holmes or to the events of his life that any sentence refers. (From (3))

5. It is not to Sherlock and or to the events of his life that any true sentence refers. (From (2) via (3)).

6. (1) is contradicted by (4), and (2) by (5). (Paradox).[31, 32]

---

[29] Frege's dismissal is named such in John Woods and Jillian Isenberg, "Psychologizing the semantics of fiction", *Methodos* online, April 2010.

[30] That is, who understand the sentences of the text and also understand that the text is a work of fiction.

[31] There is a fair question as to whether the present puzzle gives a paradox in the sense of antimony or in a looser sense of surprising befuddlement. For present purposes, it is unnecessary to settle the matter.

[32] It is true that sometimes inconsistencies occur within stories, not inadvertently but with authorial intent. There is a Ray Bradbury story in which a protagonist Keith is elected president in 2053 and it is not the case that he was. In Graham Priest's "Sylvan's Box" — in which the author indulges himself with a guest-appearance — inconsistencies are "essential to the plot." See "Sylvan's box: A short story and ten morals", *Notre Dame Journal of Formal Logic*, 38 (1997), 573-582; 579-580. Djaich da Bloo's "The Mischief of Ricardo Bosque" has its protagonist cause it to be the case that every sentence and its negation true. (This is reprinted as the appendix to chapter 6 of my *Paradox and Paraconsistency: Conflict Resolution in the Abstract Sciences*, Cambridge: Cambridge University Press, 2003; 226-227). Inconsistencies internal to stories require a different treatment than the inconsistencies on display in the first few lines of the paradox called for. For want of space, I'll leave this for another time.

Left undealt with, the paradox of fiction is a considerable annoyance to the going theories of reference, truth, and existence – a turbulent disarrangement of mainstream philosophies of language. To date, the paradox has drawn one or other of three kinds of response. One is to blow it off without ado. Another is to blow it off with ado, that is, with some kind of explanation. A third is to dissolve it in a purpose-built semantics for fiction or in some other way.

I will take it as given that simple unvoiced neglect is not an intellectually grown-up way to handle this puzzle. The second is more interesting, not least because it was espoused by Frege. But the "ado" Frege provides by way of explanation is not very convincing. A more promising option is to try for the collapse of the paradox by inconsistency-denial. I am going to argue that the harder it is to find a principled basis for inconsistency-denial, the easier it will be to look favourably on a different way of dissolving the paradox, which would be effected by inconsistency-*affirmation* coupled with *paradox*-denial. In order to secure this option, it will be prudent to proceed somewhat indirectly. So let us move now to a consideration of a semantics for fiction.

2.2   Non-entities and nonesuches

There are lots of present-day semanticists who prefer something like our second or Fregean option, that is, to locate their treatment of "the flatmate of John H. Watson, MD" and "Sherlock Holmes" in a more general framework for "empty" singular terms. This would be one in which "the flatmate of John H. Watson, MD" would be treated in just the way that "the present king of France"is, and "Holmes" in the way that "Main Street, USA" is.[33] How such a view could find such favour betrays a striking disregard for plain facts plainly on view. One of them is that fiction *moves* us. Sometimes it moves us to tears. On the other hand, it is clear that there isn't any kind of emotional reaction to the baldness of the present king of France[34], whereas emotional response to the goings-on of Holmes is a matter of course. For one thing, Agatha Christie admired him, but more generally, his readers find him *interesting*. Different examples yield different responses. Consider the horror induced by Bill Sikes' sadistic slaughter of his street-walking girlfriend Nancy.[35] We need a label for this. Let's say that the putative but nonexistent objects to which emotional response is impossible are "nonesuches", and that the putative

---

[33]See, for example, R.M. Sainsbury, *Reference Without Referents*, New York: Oxford University Press, 2005; paperback version 2007.

[34]Or, varying the example, of the wrenching despair of the present king of France, putatively reported by "The present king of France is awash in despair."

[35]As chronicled by Charles Dickens in *Oliver Twist* (1838).

but non-existent objects to which emotional response is possible are "nonentities".[36]

The same seems to be true for reference. Reference is impossible for nonesuches. There is nothing whatever of which "The present king of France" is true. But reference to nonentities is routinely achieved without breaking a sweat. Similarly for belief. There are no (positive and categorical) beliefs to be had about the present king of France, whereas there are legions of them about Holmes and Nancy. Millions believe that Holmes was a master problem-solver and millions believe that Nancy had a heart of gold.

Consider our shock at Nancy's death and our shock at JFK's death. Our shock at Nancy's death is attended by the belief that there was no such death, and is unmolested by it. But a similar belief about JFK's death could not *but* molest the shock it gave us. It is causally impossible to be shattered by JFK's death if we concurrently, steadfastly, and unequivocally believe that it didn't occur.

As we begin to see, this "double-aspectness" of fictional experience is a matter of the duality of causal pathways. The shock of Nancy's death is causally induced by what the story tells. The belief that there *was* no death is occasioned not by what the story tells, but by what the world tells. And in this instance what the world tells is that what's happened in fiction hasn't happened at all. This is a fact that Dickens knows and exploits. He could not produce his fictions without a clear-eyed awareness that none of it is so and that all of his readers will know this. But it is not part of any story of Dickens' that none of this is happening.

It is natural to think that these nonesuch-nonentity asymmetries are linked. No one – save some smattering of philosophers – seriously doubts the *affective* asymmetry. Why doubt that it is an asymmetry which so naturally adapts to doxastic reinterpretation? Of course, it is either semantically significant that the affective asymmetry is preserved under doxastic reinterpretation, or it is not. If it is, nonesuch-sentences and nonentity-sentences require asymmetrical semantic treatment. Their semantics must differ in kind.

Part of what seems to trouble Mark Sainsbury and others is that so long as they cleave tightly to lines (3), (4) and (5) of the paradox of fiction, they can't find a principled basis for differential semantic treatments of "empty" terms. I would suggest that, contrary to how things might appear to these colleagues, there is indeed a sound basis for asymmetrical treatment. It lies in the plain fact that nonentities can make us cry and nonesuches cannot. And that, it may be said, is itself a fact that rises to the bar of doxastic significance and semantic significance. How can you

---

[36] John Woods, *The Logic of Fiction*, second edition, volume 23 of Studies in Logic, London: College Publications, 2009. Originally published with the subtitle *A Philosophical Sounding of Deviant Logic* by Mouton, the Hague and Paris, 1974; p. 29 *et passim*.

know that it is her dreadful death that's caused your distress without being aware that it is Nancy whose death it was? How can you think that Nancy's death is the cause of your distress if Nancy didn't in fact die? Indeed, how can it not be the case that the person who reads his novels and short stories with understanding is a well-performing three-grade manager of inconsistency robustness? Against this it is commonly urged that we could accommodate these strong impressions of semantic success by exercising the inconsistency-denial option. If it could be successfully negotiated, then the question of the management of grade-three inconsistency robustness would be cut off at the knee. So now the question is one of how inconsistency-denial would *be* achieved and whether there is any principled basis for it other than our desire to avoid consistency.

2.3   Ambiguity

It is very widely believed by those who take an interest in them that the inconsistencies of fiction are inconsistencies in appearance only, made so by the systematic ambiguity of fiction discourse. If the ambiguity thesis were true, it would deprive me of the primary datum for the thesis that inconsistency-robustness has a much wider provenance than in software engineering. It will take this section and the next three to have my say about ambiguity. But, in the spirit of cards on the table, let me lay down two of them now.

1. Ambiguities aren't free for the asking.

2. When an ambiguity is invoked to help solve a theoretical problem, it is intellectually troublesome not to have had at hand an independent theoretical means of establishing its presence. Anything else is at risk of special pleading.

Anyone who reads or listens to a fictional story with understanding, is liberally possessed of ready beliefs about and affective responses to what the story tells. This is not unlike someone's reading news reports filed from Kiev. But unlike the Kiev story, everyone realizes that none of what is (fictively) reported in fiction has actually happened and there is nothing to which it might have happened. This commonplace occurrence occasions numberless belief-pairs roughly in the form $\ulcorner \Phi$ is the case$\urcorner$ and $\ulcorner \sim \Phi$ is the case$\urcorner$. Let's repeat the point that it is equally a commonplace that when these conflicting beliefs are transparently on view, these aren't attended by the slightest degree of cognitive dissonance. This raises a question of central importance: Why do such conflicts create such distress when they arise from set theory, yet none at all when they arise from fiction?[37]

---

[37] Here again the old prejudice recurs: "Fiction is just horsing around. Set theory is *serious!*"

A standard move – perhaps the one most commonly advanced – is to enter a plea of *ambiguity* – an inconsistency-denial manoeuvre, hence a way of dissolving the paradox of fiction.The conflicting sentences of set theory aren't ambiguous, whereas the conflicting sentences of fiction are systematically so. The trouble is that it is not at all plausible, either in common linguistic practice or in any of the known theories of meaning, that "Holmes lived in Baker Street" actually *is* ambiguous. It contains no ambiguous word and there is no ready indication of syntactic ambiguity in the manner of "Visiting relatives can be boring". What, then, is the basis of the ambiguity claim?

In relation to fiction, the ambiguity thesis has been around for at least forty years. One of the more prominent forms of it is the normally tacit invocation of a fictive-operator f, prefixed to a fictional Φ but (typically) not to its negation.[38] Also quite old is the suggestion that the f-prefix is profitably treated as a modal operator.[39] But the invocation of f as sentence-operator, with or without modal import, changes the question without answering it. The question now is: On what independent basis do we say that the sentences of fiction actually carry this operator? And, if they do carry it what makes it the case that this is a modally loaded operator if it indeed is one?

In pursuing these questions it is necessary to recall to mind the primary datum for our project, the most consequential of the plain facts plainly observable on the ground. It is not only that everyone on earth believes some variation of (1)–(5), but that virtually no one experiences the slightest difficulty in believing them jointly, or any degree of cognitive turbulence in so doing.[40] Against this it will be argued as follows: If the readers of fiction did indeed implement (1)–(5) without an iota of cognitive dissonance, that would show the inconsistencies involved to be only apparent. Why? Because people *can't* believe inconsistencies without cognitive dissonance. But this begs the very question at issue. Add to that the further fact that we are having a hard time in finding a suitably grounded case for executing the ambiguity manoeuvre, we find ourselves moving more readily to the fact that hardly any of those who embrace (1)–(5) with cognitive serenity suffers the least distress in also believing (6). So I think that we may say that virtually every human being on earth in matters such as these is a capable *two-grade* manager of robust

---

[38] John Woods, "Fictionality and the logic of relations", *Southern Journal of Philosophy*, 7 (1969), 51-64.

[39] See again my *Logic of Fiction*, 131-144.

[40] When I say that people believe (1) – (5) without a trace of cognitive dissonance, I don't intend it that, if pressed to explain *why* they aren't cognitively stressed, they might not then and there find themselves at a bit of a loss. But it's the *question* that disturbs them, not the state of mind they were in that occasioned it.

inconsistency. He is that, I say, in the absence of independently credible grounds for seeing "Bill slaughtered Nancy" as an ambiguous sentence of English. I will stay with this point a bit longer, before moving on to the possibility of full-blown three-grade inconsistency-management.

2.4 Other sources of ambiguity?

If we adopted the idea of grade two inconsistency-management, we would have the beginnings of an answer to the present question, *without the need of ambiguity*. It is that the sets of yore don't respond to two-grade management and Holmes does. The inconsistency of naïve set theory was attended, and still is, by a considerable and abiding cognitive dissonance, and there is almost nobody who actually believes that there is a set that is and isn't its own member. Neither of these conditions is true of Holmes. Nearly everyone believes that Holmes lived in London and that it's not the case that he did, and hardly any of them is in the least way disturbed by it. The difference, then, is one of inconsistency robustness. Holmes' inconsistencies are robust. The inconsistency of naïve sets is said to be as fragile as spun sugar. (I will come back to this point. Consider this a red flag.)

When I say that it is an empirical fact that virtually everyone appears to believe without cognitive distress some variant of "Holmes lived in London" and "It is not the case that he did", I am not suggesting that these same people would have, if asked, a ready or untroubled explanation of how this grade two serenity is achieved or what shows this to be a cognitively "justified" state to be in. But this is not something that should surprise (or alarm) us. Concerning virtually any of the operations of the doxastic machinery of a human individual, there is scarcely a challenge of this sort that is easy to answer. It is one thing to know that you're in a given epistemically comfortable state and another to know the ins and outs of how you got there, and from whence it derives its comfort.

Even so, some people simply won't give up on ambiguity. Perhaps they would have better luck with something that we could call "contextual meaning", itself a form of speaker meaning (I suppose). Let's grant that "Holmes lived in Baker Street" is neither lexically nor syntactically ambiguous. Imagine a context in which Harry and Sarah disagree about where Holmes lived, Baker Street or Bayswater Street. Harry opts for "Holmes lived in Bayswater Street". Harry is wrong. "Holmes lived in Bayswater Street" is false. Sarah's choice is Baker Street. Sarah is right. "Holmes lived in Baker Street" is true.

Suppose now a different context. Heckle and Jeckle are employed by city council to revise and update London's registry of residential occupancy. They are presently investigating Baker Street records from 1830 to 2003. The question is whether

Holmes should be placed in the registry. Jeckle says yes; Heckle says no. Heckle is right. "Holmes lived in Baker Street" is false.

It is an interesting case. It invites us to sort out whatever it is, if anything, about "Holmes lived in Baker Street" that makes it true in the first of these instances and false in the other. We might find ourselves thinking that there is nothing whatever about "Holmes lived in Baker Street" which *in and of itself* gets it to be the case that it carries different truth values in different contexts of use. If so, it would be hard to credit the ambiguity thesis. If "Holmes lived in Baker Street" is indeed ambiguous, ambiguity is a property of it – a property it has in and of itself. But if the context-sensitive variability of truth value is independent of any property possessed in and of itself by "Holmes lived in Baker Street" we have it trivially that ambiguity is not one of its properties.

Against this it might be argued that there certainly is some property or other in virtue of which these contextually sensitive variabilities occur. Suppose we wanted to make an inventory of the facts that constitute the life-history of Holmes. Wouldn't "Holmes lived in Baker Street" record one of those facts? Suppose, on the other hand, we wanted an inventory of all the facts that sum to the whole history of London. Would there be the slightest inclination to include the fact reported by "Holmes lived in Baker Street"?[41] It would appear, then, that "Holmes lived in Baker Street" possesses two properties that might jointly account for the alethic variabilities at hand. One is that "Holmes lived in Baker Street" is *history-constitutive* of Holmes. The other is that it is *not* history-constitutive of London. This too has come to be known as the asymmetry problem. (This is getting to be quite a lot of asymmetry!) One might expect the relational predicate "lives in" to be truth-preservingly symmetrical with its passive transformation, "is lived in by", the one instantiated by <Holmes, London>, the other by <London, Holmes>. But that is not the case here. At least that is not *quite* the case. It is not necessary to abandon the equivalence of "lives-in" and its passive transform. It is a truth-preserving equivalence, but not a history constitution-preserving one.

The present idea has been around for a long time, originating in the late 1960s.[42] It has been given sporadic but generally scant notice ever since, apart from some small pockets of favour.[43] But little has been done by way of systematic develop-

---

[41] Of course, the inventory would include that London was *fictionalized* as the place in which Holmes resided. It is true of London that it was thus used. It is not true of London that Holmes lived there.

[42] Woods "Fictionality and the logic of relations", 42-55

[43] See Nicholas Griffin, "Through the Woods to Meinong's Jungle", in Kent A. Peacock and Andrew D. Irvine, editors, *Mistakes of Reason: Essays in Honour of John Woods*, pages 16-32, Toronto: University of Toronto Press, 2005; pp. 24-26. See also Woods' "Respondeo", pages 103-

ment. The point in mentioning it here is the further discouragement of the idea that our alethic variabilities require "Holmes lived in Baker Street" to be ambiguous. While lending some explanation of the alethic variabilities, there is nothing in the history-constitution properties that plausibly requires the imputed ambiguity.[44]

2.5 Truth conditions

It is an article of faith among philosophers of language who have cut their teeth on the model theory of formalized languages that meaning is a matter of truth conditions. It is a tight connection (whatever the other details): Sentences that differ in truth conditions differ in meaning. If this is right, the semantic asymmetry is now well-established for the nonesuch-nonentity duality, never mind what we think of the nonesuch-nonentity, affective-doxastic and truth-history constitutiveness *asymmetries*. And it is established without any need of the *f*-operator. Thus "Holmes waved our strange visitor into a chair" is true because it appears word for word in one of the Holmes' stories, and "It is not the case that the case that Holmes waved the stranger into a chair" is not true because it doesn't occur in the story.[45] But the first sentence is also false and its negation true because the world makes the negation true and, negation being what it is, makes the first false as well.

This, too, is all quite rough, but perhaps it conveys well enough the general idea. "Homes waved our strange visitor into a chair" is made true by the world *in respect of* its occurrence in the story. Its negation is made true by the world *in respect of* its non-occurrence there. If meanings are truth conditions, then the nonesuch-nonentity duality is semantically significant and securable independently of the affective-doxastic dualities. Indeed, we might even say that the imputed dependencies flow backwards, with the semantic duality now taking precedence, and the others trailing along after suitable adaptation. But these latter were the very dualities from which we might seek guidance in setting up a purpose-built semantics

---

108 of the same volume; p. 105. See, in this regard, Griffin's foreword to the second edition of *The Logic of Fiction,* footnote 4.

[44]Besides, it doesn't actually seem to stand up that history-constitution preservation is lost after all. If on the basis of what the stories say is it true that Holmes lived in London, then that is as much a fact about London as about Holmes. Or, as we say, it is history-constitutive both ways in the story. Similarly, based on the fact that it is not the case that Holmes lived in London, then it is history-constitutive of neither that London is where he lived. Because he didn't; he only did in the story. Let us also note that even if the notion of history-constitutivity gave us what we wanted, it wouldn't be in the least obvious that the sentences "Holmes lived in London" and "London was lived in by Holmes" are in any independently establishable sense *ambiguous.*

[45]The scope of the clause "in the story" is still a matter of lively contention. I think we needn't settle that matter now, but interested readers might again consult "Psychologizing the semantics of fiction".

for fiction. So that motivation would appear to be in some doubt if the semantic duality could be established truth conditionally.

For these and other reasons, it matters whether the truth conditional approach is right for English. I can't see that it is. The meanings of sentences in English derive at least in significant part from their lexical meanings and syntactic organization. Neither of those ingredients is available to purely formal languages. The sentences of e.g. classical first order languages are entirely without propositional content. They lack all capacity to say something.[46] Intuitively, what they lack is meaning. But if for theoretical reasons we must somehow have it that these empty strings have meanings *of a sort*, perhaps it is not wholly unnatural to suppose that truth conditions would be the way to go, Hilbert's reservations notwithstanding. English, like every other natural language, is another story. An English sentence means what it does, and its occurrence in the sentence that negates it means the same. By the same token in, "'Holmes lived in Baker Street' is true" and "'Holmes lived in Baker Street'", "Holmes lived in Baker Street", means the same each time. So sameness of meaning can override different truth conditions.

If at this point we decided to grant some standing to the thesis of two-grade inconsistency management, we could venture an opinion about the source of inconsistency's *value*. The short answer is that its value is a comparative one, *faute de mieux,* made so by the impossibility of keeping it out or, once in, getting rid of it in any convincing way. *It is the value of not having to draw blood from a turnip.*

I would not wish to leave the impression that I think truth conditions play no role in the current treatment of fiction. The opposite is true. Truth conditions are essential to my examination of the three-grade management thesis. What is not essential, and what is not true either, is that truth conditions are what confers meaning on the statement-making sentences of natural language. The particular importance of this is that the truth-value difference between "Sherlock lived in London" and "It's not the case that he did" can be accounted for without the necessity to invoke or pretend into existence the ambiguity of "Sherlock lived in London".

It cannot be denied that lots of people simply won't let go of the point that fiction is such small beer in human cognitive life that, even if what I've said of it here were so, it would offer the IRL machinery only the scantest promise of applicational fit to the human condition as played out *in terra firma*. I would like to rebut that notion

---

[46]Consider the simple case of propositional logic. The meanings that formal sentences of the propositional calculus have are contributed by the meanings conferred on their connectives by way of their conditional truth functional contextual definitions. The meanings that are conferred on sentences are conferred by arbitrary assignment of truth values to atoms regulated in turn by truth-functional interpretations of connectives for molecular sentences. But nowhere in the process does it get to be the case that a formal sentence *says* anything.

now. Stories are governed by an *anti-closed world* axiom.[47] Stories inherit the world as it actually is, in all particulars save for the deviations imposed by the author. If this weren't the case, stories would be so awash in indeterminancy, so deprived of background conditions as to be virtually unreadable. Boredom kills readership. Utter boredom kills it utterly. So not only does Holmes have an aliamentary canal, he does or does not have a mole on his backside. Not only does Holmes live in London, he lives in a place a few thousand miles east of Moose Jaw, and a considerable distance west of Samarkand.[48]

This matters. Let R be a relation between a fictional being and some non-fictional object in the world. Then anything P that is non-fictively true of the non-fictional relatum gives something to which the fictional relatum bears R or some variation of it: If Holmes bears R to London, he also bears it to a place which has all those P. If Holmes lived in London, then Holmes is implicated in everything in the London's spacetime worldline. Every sentence in the ambit of those of the form that give the first five lines of paradox of fiction has this same involvement. Each constitutes a *worldline-inconsistency virus*. The cardinality of these relational facts is probably finite, but not in any event anywhere close to small.

Still, we now need a further distinction between real world truths that are true *in* a story and real world truths that are *part* of a story. That the real numbers are undenumerably many is a real fact. Hence, by the anti-closed world axiom, it is true in the story. But no one seriously thinks that it is any part of the story told by *The Hound of the Baskervilles*, any more than it would be part of the reporter's account of today's fraught doings in Aleppo. So not everything that's true in the world is made *false* by its also being true in the story.

My task here is to see to what uses the Hewittian notion of inconsistency-robustness might be put beyond the reaches of the metamathematics of formal systems. To that end I have developed the idea that inconsistency robustness comes in different grades, the higher the grade the more philosophically interesting and contentious. A key part of Hewitt's concept is that inconsistency-robustness is an aid to the *management* of large information systems. To that end my further purpose is to consider how the manager of the information system of a human individual living in a large community of his fellows might proceed with handling each grade

---

[47] A referee asks why we wouldn't consider adapting the many worlds interpretation of quantum phenomena for the semantic analysis of fiction. My answer is that the semantics of fiction requires no worlds other than the one we're all in.

[48] "Wait!", I can hear being cried from near and afar, "how can you say that Holmes lived in a place quite a bit west of Samarkand? To say this, you need the intersubstitutativity rule which, fictions being intensional contexts, is unavailable here" To which I reply: The rule's invalidity in intensional contexts does *not* deny it successful application in all cases. It denies it successful application in *at least one* case.

of his system's inconsistency robustness (hence the question of three-grade management.) In my submission so far, I have put it that relative to design and capacity, the human agent is (or runs) a large information system, and that a large subset of it is robustly inconsistent at each grade. Accordingly, the system itself is robustly inconsistent at all three levels.

Pervasive inconsistencies give a rather nice sort of partition between how they are handled in everyday belief-forming practice and how they are viewed by belief-management specialists. It is not, however, a completely steadfast partition. As mentioned earlier, one way to disturb the repose of a person's ready embrace of inconsistency is to get a belief-management theorist to press him with the "How dare you!"-challenge. I have already suggested that when pressed, the everyday believer has neither the information nor the skill to say with authority how it comes to pass that the inconsistencies of fiction are both widely recognized and widely and readily believed. But I want to say that when, under this pressure to theorize, the layperson might himself offer up the suggestion that the inconsistencies he's so at home with aren't really inconsistencies after all. My response to that would be that the layperson has got it wrong.

At the root of it all is the embedded certainty that the recognized inconsistencies that don't bother us are inconsistencies that we should be bothered about, that a failure to bother is somehow a failure of our cognitive integrity. Since virtually everyone on earth is unbothered by the manifest inconsistences of fiction, humanity stands convicted of a rather hefty belief-management ineptitude. Many will believe – some of them the very people mentioned in lines above – that self-denunciation on this scale is more than the evidence will bear. They will seek a way to ease its burden. By far the most favoured remedy is inconsistency-denial. The inconsistencies that don't bother us aren't inconsistent after all; their appearance of inconsistency is false. Anyone drawn to this remedy has a stake in answering the central question it raises. What *makes it the case* that these appearances are deceiving? Here, too, there is a dominantly arrived at answer. The apparent inconsistencies of fiction arise from ambiguity.

In so saying, several other questions come tumbling in. Two of the most important are: To *what* do these ambiguities attach? And, apart from a theorist's eagerness to avoid a doxastically relaxed response to inconsistency, what *justifies* these ascriptions to those parts? Is there a known theory of ambiguity that makes independent provision for them? I have been arguing here that to the extent that the burden of answering the first question has been met, the burden of answering the second hasn't been. Question one has attracted a number of candidates – lexical ambiguity, syntactic ambiguity, contextual ambiguity, truth conditional ambiguity – for which (I say) no adequate answer to question two has succeeded, even when offered. The more that this is so, the less good the ambiguity thesis seems as an implementation of the inconsistency-denial option.

2.6 Speaker meaning

When I say that it's true, not false, that Holmes lived in London, I don't mean to say that it is true in actuality that Holmes lived in London. Neither do I expect that this is how you will take what I've said. I expect that you will take from what I've said that it's true in the story that Holmes lived in London. Everyone who knows Holmes – that is, has read with understanding the stories about him or who has been told of him by someone else who has read with understanding his stories – will know that what makes it true that Holmes lived in London is Doyle's stories to that effect.

When these conditions obtain let's for convenience speak of speaker and listener as "persons of understanding". When a person of understanding speaks to a person of understanding the words "Holmes lived in London", he is exploiting the pre-existing shared belief that words have stories as their truth-makers.When, as someone of understanding, I speak these words to another person, I would not use *these* words unless I thought that he too was a person of understanding. If I didn't think so, I'd have used different words, something like, "Sherlock Holmes, the fictional detective, lived in London". My addressee would now himself be a person of understanding. Suppose a third party, also a person of understanding, is a philosopher of language with a keen interest in theories of truth for natural language. He asks, "What does 'true' mean here?" If you answered, "It means 'true in the story'", your answer would be false on its face. It would confuse a truth with a truth-maker, which was a long-discredited mistake of some of the old positivists. There is a further difficulty with this. In any context in which "true" means "true in the story", the latter can contain no occurrence of the former. Otherwise, we would have it that "Φ is true in the story" is true in the story. But it isn't, so we don't; and since we don't, "true" on those occasions can't mean "true in the story". I suppose something might be said for the suggestion that when a fictional sentence Φ is true, it is also true that "Φ is true in actuality" is true in the story. But do we really want to say that when a person of understanding speaks the truth of Φ that "true" means "true in actuality in the story"?

Some people will think that I've fallen into a confusion occasioned by an inadequate grasp of the distinction between lexico-sentential meaning and speaker meaning. They will allow that the word "true" means here whatever it may mean elsewhere, as in "It is true that Mayor Ford lives in Toronto". The difference lies in what the speaker means now and what he meant elsewhere. The former is made true by the story, the latter by Mr. Ford himself. When someone tells me that when I say Φ what I mean to say is Φ^*, I always ask him back, "Well if that's what I meant to say, why didn't I say *it*?" Perhaps your rationale on my behalf is that Φ would be an efficient paraphrase of Φ^*. Perhaps it is. But it remains puzzling that if in uttering Φ what I meant to *say* was its paraphrastic converse Φ^*. No, what I meant to say was Φ, and the reason I wanted to say *it* is precisely because it paraphrases Φ^*. *That* is the fact, if fact it be, that motivates the saying of Φ rather than Φ^*.

Of course, "what I meant to say" carries ambiguities of its own. Sometimes "say" means "utter" and at other times it means "convey". Perhaps a more plausible suggestion is that when a person of understanding says "Holmes lived in London", what he means to convey is that Holmes lived in London in the story. But notice that here too if you expressed the proposition you wanted to be conveyed using the words "Holmes lived in London in the story", it can hardly be the case that what you intended to convey by the occurrence therein of "Holmes lived in the story" is that Holmes lived in London in the story. For then we would have it that when you say that Holmes lived in the story what you intended to convey was that Holmes lived in London in the story in the story. But it isn't true; so that's not what you intended to convey by "Holmes lived in London". The truth of the matter seems to lie less distantly and arcanely at hand. When a person of understanding says that London is where Holmes lived, he intends to exploit a shared belief between persons of understanding. It is a belief about the truth-maker of the thing said, which is said with the expectation that its hearer will know how to assess its accuracy. What we have here, then, is expectation, not conveyance. Still, there is something to be said about conveyance. If speaker and hearer share this belief about the truth-maker, there must have been something or other that conveyed it to them. The answer lies in the idea of a person of understanding and how the idea is instantiated. In the simplest kind of case, such as the example currently in view, it is a shared appreciation of who Holmes is, of Holmes' standing as a person of fiction. In a great many cases, there is a shared prior acquaintance with the workings of "Holmes". In other cases, prior indication of its specialness is given in a context-setting way. "Let's talk about London" is not one of them. "Let's talk about fictional detectives" usually is.

There is no cause for surprise in any of this. It is entirely commonplace for assertive utterance in general to carry expectations of a shared engagement of background information. However, it is hardly ever the case that such expectations are announced by the utterer's words or by whatever he may have meant by them.[49]

---

[49]A referee points out that on p. 29 of *Reference and Existence*, New York: Oxford University Press, 2013, Kripke proposes that names that occur in fiction aren't names at all, but only "pretended names". It is true that pretence theories of fiction are one way of achieving inconsistency-

2.7 Of and in

Let me say again that I think that the "small beer" accusation should be resisted. The inconsistency robustness glass contains a good deal more of the beer of fiction than we might at first glance suppose. Stories are governed by an anti-closed world axiom. Stories inherit the world as it actually is, in all particulars save for the deviations imposed by the author's own provisions. If this weren't so, stories would be so awash in indeterminancy, so deprived of background conditions as to be virtually unreadable. Boredom kills readership. Utter boredom kills it utterly. So not only does Holmes have an aliamentary canal, he does or does not have a mole on his backside. Not only does Holmes live in London, he lives in a place a few thousand miles east of Moose Jaw, and a considerable distance west of Samarkand.

This matters. Let R be a relation between a fictional being and some non-fictional object in the world. Then anything P that is non-fictively true of the non-fictional relatum gives something to which the fictional relatum bears R or some variation of it: If Holmes bears R to London, he also bears it to a place which has all those P. If Holmes lived in London, then Holmes is implicated in everything in the London's spacetime worldline. Every sentence in the ambit of those of the form that give the first five lines of paradox of fiction has this same involvement. Each constitutes a *worldline-inconsistency virus*. The cardinality of these relational facts is probably finite, but not in any event anywhere close to small.

Still, we now need a further distinction between real world truths that are true *in* a story and real world truths that are *part* of a story. That the real numbers are indemumerably many is a real fact. Hence, by the anti-closed world axiom, it is true in the story. But no one seriously thinks that it is any part of the story told by *The Hound of the Baskervilles*, any more than it would be part of the reporter's account of today's fraught doings in Kiev. So not everything that's true in the world is made *false* by its also being true in the story.

2.8 A brief look at pluralism

This brings to a close my review of the ambiguation thesis in all its principal contemporary variations. Perhaps some readers will think that the wait has been indefensibly long. I respect the sentiment, but am unable to bring myself to share it. My primary datum here is fiction, made so by the empirically discernible fact that, for anyone who reads or hears stories, there is a pervasively inconsistent set of beliefs

---

denial, but I'm at a loss to see any empirical support for the idea that "Holmes", "Sykes" and the others aren't names at all. Equally, the idea that we when we read stories we are having only a pretend engagement with its goings-on also lacks empirical warrant. When we cry over Nancy's death are our tears only pretend tears, and not tears at all?

that he holds without a shred of cognitive anxiety. This, if true, is inconsistency-robustness with a happy face. The dominant theoretical position on the inconsistencies of fiction is that they don't exist, that they are misappearances occasioned by equivocation. If the inconsistency-denial position held true, I would lose what I take to be the single-best exemplar in the cognitive life of the human animal of an inconsistent system as robust as it gets. The ambiguation strategy is by no means limited to the semantics of fiction. It has a busy life of its own, and is lavished upon problems in which inconsistency has only a limited role, if any at all. Nowhere is this more evident to logicians than in the pluralist response to logic's own sprawling and often rivalrous multiplicities. Take the modal logics of propositions as a quick example, beginning with the original five systems of Lewis as set forth in 1932.[50] Within the next three decades there were something on the order of fifty different propositional systems, each an extension or restriction or adaptation of Lewis' five, that had made the scene in modal logic.[51] On the face of it, this is promiscuity run amok, an *embarrass de richesse* that beggars belief. How could a common subject-matter endure so rivalrous a plentitude? One of the standard answers calls upon ambiguity. There are two versions of it. In one, for every fifty different treatments of $\Box$ and $\Diamond$, there is a correspondingly different meaning of "necessarily" and "possibly". The other has it that even if there weren't fifty different such meanings antecedently in play in English, it lies in the logician's remit to make new meanings up. I lack the space to examine these options here. Suffice it to say that I find the first one absurd on its face, and the second a threat to motivational coherence.[52] Concerning the first, there is no known theory of ambiguity in which "necessarily" is fifty-wise ambiguous. Concerning the second, what we might well have is a new meaning for $\Box$, but what we *won't* have, short of lexical theft, is a new meaning for "necessarily".

The point of it all? Invocations of ambiguities as a problem-solver considerably outreach their availability. Ambiguity is mismanaged in the pluralism of modal logic. It is also mismanaged in the logic of fiction.

2.9 The law of noncontradiction

It now falls to us to consider two main options. One is to find a way to deny inconsistency without the pretense of ambiguity. The other is to find a way to

---

[50]C.I. Lewis and C.H. Langford, *Symbolic Logic,* New York: Appleton-Century Croft, 1932; reissued by Dover, 1956.

[51]Roughly and compactly speaking, a system makes the scene in logic when it issues forth in the pages of the *JSL*.

[52]For more readers could consult JC Beall and Greg Restall, *Logical Pluralism,* New York: Oxford University Press, 2006, and again my "Inconsistency: Its present impacts and future prospects".

affirm the inconsistencies without disturbing the repose of those who believe them so readily. Speaking for myself, the second is worth a try. Perhaps the notion of *respects* will play a helpful role. Perhaps it will prove to be critical.

Perhaps some readers won't much like the suggestion that truth-makers for "Holmes waved our strange visitor into a chair" and its negation are both provided by the world, albeit in different respects. A world that provided truth-makers for sentences and their negations would in turn be inconsistent, and surely that would violate logic's oldest and most secure principle, the law of noncontradiction. I have two things to say about this. One is that the world's inconsistency is precisely what I'm presently inviting a patient consideration of; so it is dialectically unavailing to withhold that consideration on grounds of what it would be a consideration *of*. The other is that even if the world were inconsistent, it is an open question as to how that would stand in regard to the law of non-contradiction (LNC).

LNC is arguably Aristotle's single best-known idea. In the *Metaphysics* he gives it three different and pairwise inequivalent formulations, and does so in the following order:

- *The doxastic formulation:* No one can believe that the same thing can (at the same time) be and not be. (1005^b 13-14)

- *The ontological formulation:* It is impossible that the same thing belong and not belong to the same thing at the same time. And in the same respect. (1005^b 19–20)[53]

- *The logical formulation:* The most certain of all basic principles is that contradictory propositions are not true simultaneously. (1011^b 13-14)

Of the three the ontological gives the fullest formulation of Aristotle's thinking, which means that the other two are incomplete.[54] This is important: The other

---

[53] Cf. Plato: "The same thing clearly cannot act or be acted upon in the same part of relation to the same thing at the same time in contrary ways." (*Republic*, 436B)

[54] Perhaps the ontological formulation is itself somewhat lacking. Immediately after the words quoted here, Aristotle goes on to observe that "We must presuppose, in the face of dialectical objections, any further qualifications which might be added." An explanation of this puzzling remark is ventured in chapter 5 of my *Aristotle's Earlier Logic*, soon to be reissued in a revised edition by College Publications of London; it first appeared with Hermes Science in 2001. There is a further question now to consider. Aristotle says that a first principle is a proposition that is true, necessary and "most intelligible", and it neither requires nor admits of proof. If, as Aristotle avers, LNC is the most certain of first principles, then it is a principle that neither requires nor admits of proof. But consider the following argument:

i. LNC is a first principle. By agreement.

two are incomplete. In particular, the *logical* formulation understates the law.[55] It is here that we might find the ontological formulation and the third-grade of inconsistency robustness of being doing one another some good.

Let's come back now to the Russell inconsistency. There is something to be said for the view that it does indeed do violence to Aristotle's law. Most people appear to think that the world does not make it true that the Russell set both is and isn't its own member. Of course, by the axioms of naïve set theory this is precisely what the world does do. They provide that it lies in the very nature of sets that some will and won't be their own members, don't they? No. This is what the axioms *say* is true of sets. But what the axioms say is false, not true. There is no one respect in which they are and some different respect in which they aren't, each of which is a truth-maker for sets. Fiction is different.

It is taken as plain fact that the damage to LNC done by the Russell paradox is attended by high levels of cognitive dissonance. It is taken as plain fact that the damage done by the Holmes-sentences does no such thing. Inconsistency is by turns an effector of turbulence and the preserver of serenity. This gives a further duality to take note of and, if we can, explain.

Perhaps something like this will do. The Holmes inconsistencies are robust. The Russell inconsistency is more fragile than spun sugar. The Russell inconsistency ruptures LNC. Holmes' inconsistency does no such thing. Human beings know this distinction without tutelage. Their awareness is implicit and subconscious.

---

ii. LNC is true. From (i) by def.

iii. LNC cannot be proved. From (ii) by def.

But ⟨(i), (ii), (iii)⟩ is a valid argument with true premises. It is a proof of LNC at line (ii) and a proof that LNC can't be proved at line (iii). So, if our proof stands, whatever else it is LNC is not a first principle. Indeed, nothing is a first principle. Here again, however, we appear to have met with Hewittean resistance. Is our proof a proof by contradiction? If it is, perhaps first principles would be back in business if *my* Hewitt's proof turned out to be right. Proof by contradiction would have to go. A small addendum: If LNC is a necessary truth – a truth of logic – then proof theory provides that it is its own proof. Apart from the fact that no one in the wide world thinks that the proofs of proof theory are what mathematicians produce when they demonstrate the laws of arithmetic, if LNC were its own proof, it still couldn't be a first principle.

[55] Perhaps this is not all that surprising. The logical formulation is the one that holds sway in the *Prior Analytics*. It is a formulation that pivots on the technical notion of contradictoriness, whose defining characteristics are reflected in the square of opposition. In it we see that the reflexive relation of being a contradictory of is defined for the following pairs of categorical proposition schemata: {⌜All A are B⌝, ⌜Some A are not B⌝}. and {⌜No A are B⌝, ⌜Some A are B⌝}. Since the language of categorical propositions lacks the expressive capacity for respects, and the logical formulation of LNC is tailor-made for categorical propositional languages, it lacks the means to capture the ontological intent of the law. This, as I say, can hardly be surprising. Formal representations routinely understate what they are intended for.

207

Accordingly, we might consider supporting

> *The IR hypothesis*: An inconsistency has the grade-three robustness of
> being when it rings none of the cognitive dissonance alarms, and so gives
> no affront to LNC.[56]

Proof by contradiction holds up only when a contradiction violates LNC. Suppose
that Φ implies for some ψ that ψ and $\sim \psi$. If the truth of $\ulcorner \psi$ and $\sim \psi \urcorner$ discomports
with LNC, then Φ fails by proof by contradiction. On the other hand, if $\ulcorner \psi \wedge \sim \psi \urcorner$
causes LNC no grief, then proof by contradiction fails. The moral: could it be that
proofs by contradiction work only for contradictions that violate LNC? What then
of "my" Hewitt's proof of the Curry-like result for consistency? As far as I can see,
no one really believes that every theory or information system is consistent. This
suffices to extinguish inconsistency robustness not only at grade three but also at
grade two. There is no respect in which anyone thinks that the universal-consistency
result is true. The omni-consistency result violates LNC.

2.10   A touch of dialethism

My thesis of three-grade inconsistency management is a dialethic thesis.[57] It is
dialethism without tears or regret. It is dialethism on a grander scale than is found
in Australia and Brazil. The robustness of inconsistency helps us see dialethism in a
new, and I think, deeper light. The Russell inconsistency was a founding motivation
of it (though not for Russell), and the Tarski inconsistency too (though not for
Tarski). Let me deal here with the Russell.[58]

The near-universal response to it was to search for a stable new home for sets,
in effect, to patch the old theory up. Interestingly enough, these weren't the responses of Frege and Russell. Both thought that the paradox destroyed the very

---

[56]In earlier pages I said that I would advance considerations in support of the idea that in
matters of inconsistency robustness it needn't matter whether we have or lack the adjunction rule
for conjunction. The IR hypothesis says that what matters here is not whether Φ and $\ulcorner \sim \Phi \urcorner$ imply
$\ulcorner \Phi \wedge \sim \Phi \urcorner$. What matters is whether any and all conjunctions $\ulcorner \Phi \wedge \sim \Phi \urcorner$ violate the full version of
LNC. The answer is no.

[57]The logic of true contradictions precedes the name of dialethic logic. Good historical coverage is
provided in Graham Priest, Richard Routley and Jean Norman, in *Paraconsistent Logic*, 1989. Franz
Berto's *How to Sell a Contradiction*, volume 6 of Studies in Logic, London: College Publications,
2008 also has some excellent background information. Lawrence Powers' *Non-Contradiction*, volume
39 of Studies in Logic, London: College Publications, 2012, is an astute and probing overview of
the long philosophical career of LNC.

[58]For consideration of Tarski, see my *Paradox and Paraconsistency*, chapter 7, and "Dialectical
considerations on the logic of contradiction I", *Logic Journal of the IGPL*, 13 (2005), 231-260..

idea of set, that indeed there was nothing there for a retrofitted post-paradox set theory to capture. It is true that Frege did tarry awhile with the retrofitting option. But this created new difficulties and inconsistencies of its own, and Frege gave up on set theory altogether. Russell's approach was more equivocal and (some would say) cynical. He too thinks that the paradox annihilated the very concept of set. Russell (following Moore and echoing Kant) thought that the business of philosophy is the clarification of concepts already in circulation. He agreed with Frege that the paradox showed it to be philosophically impossible to clarify the concept of set. On the other hand, Russell also held that it was the job of mathematicians to make up *new* ideas. (Kant thought this too). The job of philosophy is analysis, and the job of analysis is to make concepts clear. The job of mathematics is synthesis, and the job of synthesis is to make clear concepts.[59] *Qua* synthetic theorist, Russell was not out to repair old sets, but to make new ones that he hoped would prove useful to mathematics in the ways (forlornly) hoped for by the old ones. *Qua* analytic theorist, Russell had nothing to say about sets.[60] The dialethic response was audaciously different. It wasn't the retrofitting of sets that was needed. It was the upgrading of our inconsistency-robustness mechanisms, hence our cognitive dissonance-mechanisms as well, so that there could be business as usual for sets (with some compensating adjustments) but a *substantial* reconfiguration of logic.

Graham Priest speaks of this rather directly as long ago as 1987, somewhere in *In Contradiction: A Study of the Transconsistent* where he tells us that it took him a long time and considerable effort before he could look upon any proposition as both true and false.[61] In this he evidently succeeded. The question is, how did he do it?

Perhaps it just comes down to this: Priest tried and succeeded in upgrading his cognitive dissonance mechanisms from an operational condition of untutored automaticity to a more nuanced and agent-controlled facility, open to tutorial refinement. If Priest could teach himself to like true contradictions, presumably he could do the same for us. The question is, could he?

By the IR hypothesis, automatic devices fire into cognitive dissonance when an inconsistency is unrobust. Unrobust inconsistencies are inconsistencies that violate

---

[59]One might hear the ambiguation bushes now stirring. Isn't making up a new concept akin to making up a new meaning? The answer is no. When the inventors of S6 and S7 assigned different new meanings to $\Diamond$, they created new concepts, but they weren't making it the case that "possibly" has meanings heretofore recognizable or present in linguistic practice. What they were doing was high-jacking the lexicon for a word of settled usage to reflect a concept with no prior presence in linguistic practice.

[60]See here *Paradox and Paraconsistency*, pp. 141-143ff.

[61]Dordrecht: Kluwer, 1987. See also Priest, *Beyond the Limits of Thought*, Cambridge: Cambridge University Press, 1995.

LNC, Nearly everyone thinks it difficult to sell the idea that the Russell inconsistency isn't unrobust after all. Assuming that the world did indeed make the Russell contradiction true, it would be difficult to discern the appropriate difference in *respects*. Perhaps someone of priestly vocation might say that in respect of R's membership in R, R is not a member of R, and in respect of its non-membership in R, R is a member of R. So the world does after all make both propositions true but not in the same respects. LNC remains unmolested. Turbulent inconsistencies violate LNC, non-turbulent inconsistency does not. (See below, section 11). Speaking for myself, I am unable to locate this pair of respects. But I concede that this manoeuvre, if taken, would put considerable weight on the necessity to get clear about respects.

Running alongside is an answer of a kind as to how he did it, that is, how Priest learned to restrain his cognitive dissonance devices in the presence of the Russell set. The answer could be that he trained himself to be observant and to search out heretofore undiscerned respects of significance for the matter at hand. He taught himself to tease out the respects that spare the Russell the cognitive dissonance that routinely attends it. He taught himself some intellectual self-control.

I don't know that I think that any of this is a tellable story about respects. Perhaps it isn't. Perhaps it can easily be shown to be a clumsy and artless piece of sophistry. Whether tellable or not, it is certainly not the story that dialethists tell. Dialethists have no skin in the game of difference of respect, beyond its inapplicability here. Where the answer presently on view seeks the safe harbour of different respects, dialethists themselves take refuge in difference of *truth value*, notably in the difference between T and $\{T, F\}$ and F and that same pair. Accordingly, they rewrite Aristotle's law – not the more fully fitted ontological law but the already under-performing logical variation of it – which now provides that no statement can be true-only and false-only at the same time. This, they say, allows for exceptional cases in which a sentence is not T-only or F-only but is $\{T, F\}$ only. Of course, it would seem that $\{T, F\}$ implies T and does the same for F. If so, a statement that is $\{T, F\}$ is T but not T-only, and likewise is F but not F-only. But this leaves us with an unexplained difference between T-only and T and F-only and F.

We now have a new question to consider. How plausible is it to suppose that the proposed (and very peculiar) three-valuedness of the Russell statement would indeed still the dissonance classically affirmed of it and routinely triggered by it. The jury, I think, is still out. So far the indications are nothing but discouraging. Modern mathematicians seem to hate inconsistency like death itself. Until it starts to happen with some real robustness, the very idea of inconsistency-override remains in doubt.

Before closing this section, here is a last objection briefly to consider. Some readers will say: "What in the world is a respect? How are respects to be *indi-*

viduated?" It is a good and necessary question. In partial *ad hominem* reply, let me simply ask how the property of individuation is *itself* to be individuated? Let me add that hardly anything of value to the intellectual efforts of human agents is subject to sharp-elbowed individuation. More forthcomingly, a respect is a way in which the world is, a way in which the world manifests its being. If saying so restirs the demand for what inviduates worlds, I'll make do (for now) with an already settled and cherished answer to a closely related question. What, we ask, is a *possible* world? It is, we are told, a way in which the world might be or might have been. Then, before you know it, we are writing semantics for possible worlds in which ways of the world are confidently quantified over, without further ado about how *they* are to be individuated. (Their putative formal representation as abstractly set theoretic structures is another thing entirely. No one should think that it falls to a possible worlds semantics for $\ulcorner \Diamond \Phi \urcorner$ to say what a possible *world* is.).[62] For now the best I have to offer is that respects are relativized to truth makers. One and the same unambiguous sentence might be assigned T by a truth maker, and its negation, without change of meaning, be made T by a different truth maker. Since truth makers are provided by how the world actually is, respects can be taken as different ways in which the world actually is, a reflection of the complexity of its truth-making provisions. In so saying, there is plenty of room for the fact that a great many sentences — indeed an ample majority of them with unmade-up subject matters — will have truth makers and their negations will have none.

Of course, these remarks fall more into the category of plea than principled solution. It simply cannot be denied – nor should it be downplayed – that my thesis of third-grade inconsistency robustness stands or falls with the ontological reading of LNC, and that LNC in turn stands or falls with the world's respects. Until we have a more or less mature theory of respects, my grade-three position becomes less a thesis than a proposal for a research programme. With that concession now made, can I honourably retreat to the safety of the second-grade doxastic variation? I said early on that this was indeed a safe harbour, and as far as I can tell it actually is. But it offers safety with a difference. We now have a different explanation (and a weaker one) of why the inconsistencies of set theory trigger our cognitive-dissonance alarms and the inconsistencies of fiction, jurisprudence do not. The basic answer is that we believe the latter and do not believe the former. That leaves the question of *why* this would be so unanswered at the grade two level, beyond recalling to mind the semantic purport of belief.

Perhaps this is the place to pause long enough to redeem an earlier pledge. As we

---

[62]See here my "Making too much of worlds", in Guido Imaguire and Dale Jacquette, editors, *Possible Worlds: Logic, Semantics and Ontology,* pages 171-217., Munich: *Philosophia*, 2010.

saw, Frege's dismissal of fiction rested on the conviction that fiction doesn't matter for what matters to him. Logic and mathematics matter for Frege, and the last thing that can now be said with any plausibility is that fiction doesn't matter for logic and therefore for mathematics too. Frege's dismissal was an ill-considered and precipitate mistake. On the contrary, fiction tells an instructive lesson about human cognition.

2.11   Sayso semantics: A concluding promissory note

Readers will have noticed that the cases I've so far offered up for three-grade management consideration have a peculiar feature in common, and some would say a disturbing one. It is that in my treatment of fiction. I've accorded a semantic significance to *sayso*. By this I mean that I have allowed for cases in which saying that something is the case gets it to *be* the case. Another interesting source of sayso is criminal jurisprudence. When a jury says that the accused is guilty as charged that makes it a legal fact that he is indeed guilty as charged. It is not irrelevant that some notable thinkers consider the facts of jurisprudence to be fictions.[63] Of course, here too there is saying so, and there is saying so. The sayso of a story's author and the sayso of a jury's verdict differ considerably in both uptake and success conditions. But the commonality of semantic, i.e. *alethic*, significance is solidly in place. It is solidly there if the accounts I've been advancing are allowed to stand. Still, the variabilities of truth-making sayso should not be underestimated. When an author writes his stories, it is true to say that he is making things up as he pleases, and that, in so doing, he has a quite striking latitude. It is different for juries and judges. They are not making things up. Authors make things true by making things up. Juries and judges also make things true but not by making things up as they please. What they *please* has no role to play, however slight. These are large differences no doubt, but they leave the common element untrifled with. In each case, things are made true by being said to be so. This point will be lost on some people. To them I offer Russell in reply. Russell thought that the paradox cancelled the very possibility of providing a conceptual analysis of the predicate "is a set" which would specify the extension it actually and antecedently has. He was moved instead to take the route of nominal definition, by which an extension for "is a set" will be contrived by creative stipulation. It is as clear as the day is long that Russell did not for a moment think that the extension of *his* predicate for sets was null.

Even so, dissatisfactions are likely to arise. The very idea of speaking things into

---

[63]Notably Jeremy Bentham, *The Theory of Fictions,* in C.K. Ogden, editor, *Bentham's Theory of Fictions*, New York: Harcourt Brace, 1932. For developments closer to home see my "Against fictionalism".

truth will strike many people as absurd on its face. I don't think that such reactions should in the least surprise us. But before reaching a final decision, there are two further considerations that argue for consideration. One is that the idea of fictions has by now penetrated the philosophy of empirically well-confirmed model-based science. The other is that an Aristotelianly motivated theory of ontological respects – a theory attuned to the ontic formulation of LNC – allows for a relativitic account of truth without disturbing a full-blown ontic realism. Indeed the particular moral of the inconsistency robustness of being is this:

> *The realist thesis:* To be a good three-grade manager of robustness it is
> *not* required that realism be given up on. To say that the world really
> is inconsistent isn't to say that the world isn't *real.*

Let's now briefly turn to model-based science. Key to it all is that models indispensable to some of our most empirically successful sciences are serious distortions of the phenomena they are intended to model. A handy example is population genetics. It is a highly successful empirical theory of the workings of natural selection on the ground, i.e. as it actually manifests itself in nature. The solid success of the theory at the empirical checkout counter depends on a number of factors, needless to say, some of which fail utterly at the empirically checkout. By "utterly" I mean not merely approximate failures, but failures transfinitely removed from empirical accuracy. A case in point is the incorporation in the theory's model of the proposition that populations are infinitely large. Every population of whatever size allowed by nature herself falls infinitely short of infinite largeness. The infinite cardinality axiom is, as we might say, transfinitely false for all populations external to those of this idealized model of population genetics.

There are problems with this. The infinite cardinality axiom is indispensable to the impressive empirical success of population genetics, but the axiom is transfinitely false. How, then, can we think that population genetics in any degree advances our knowledge of natural selection on the ground (which is the only place it actually happens).

There is no want of answers. One is a fairly crude instrumentalism, in which. science doesn't advance our knowledge of nature, but at its best provides reliable measures for prediction and control. With instrumentalism comes the abandonment of all prospects for a realistic account of nature. No working scientist in any moment of intense concentration would think of his efforts as epistemically and ontically infertile. Instrumentalism is for the weekends, when the lab coat has been replaced by the golf shirt.

Another option is relativism. This is where model-based science stands closest comparison to fiction and serves as the spur for the fictionalisms that have arisen in the philosophy of mathematics, the philosophy of science, and in some of the more technically oriented areas of metaphysics.[64] On this approach, the infinite population postulate is true in the model but false in its application to nature. The very idea that essential components of an empirically successful theory might be both true and false sends out undeniable shockwaves.[65] It is unnecessary here to list all the complaints that arise. But there is one in particular which, at this juncture of our proceedings here, it would be prudent to focus upon. It is that the present option is disqualified from serious contention by its implied relativity of truth – truth relative to a theoretical model and truth relative to the empirical world. For isn't this relativity a killer of realism?

It is an important worry, made so by the occasion offered by our speculations of late to assuage it. The heart of the worry is that some things are simply too important – too hot – for a relativized notion of truth to handle. There is something deeply alarming in a relativistic appreciation of the holocaust – perfectly dreadful relative to your value system and perfectly fine relativized to the value system of the Nazi high command. Relativistic accommodation gives equal offence to both parties. For each is of the view that his side of the matter is the one that is true in the world. The alethic relativities of fiction, law and model-based science aren't like this. When Doyle made it true that London is where Holmes lived he didn't intend to (and didn't) enlarge the population of that great city. He was not trying to make this true in any respect that world is in, save the story in respect of which Holmes lived there. When a jury finds the odious Spike not guilty (say on technical grounds) there is no presumption that this cancels what the world has to say about the situation in respects other than the world's provision for legal fact-making. When a population biologist makes it true in his model that populations are infinitely large he has no designs on how the world provides for the matter in respects other than the provisions of his model. This is a key difference. A truth predicate relativized to respects in harmony with a cognitively non-dissonant acceptance of it. But everything pleaded as a value of x in the dyadic schema "x is true relative to y" is a respect in which the world is. There is no respect in which it is true that the holocaust was necessary and beneficial — a perfectly just bit of ethnic tidying up. The trouble with the relatisim asserted for the holocaust is the utter laziness of the relativist to specify the converse domain of his truth predicate.[66] This turns out to be an important

---

[64]See again "Against fictionalism" and, for a survey, John Woods, editor, *Fictions and Models*, Munich: Philosophia, 2010 and the ample references therein.

[65]See here John Woods and Alirio Rosales, "Virtuous distortion in model-based science", in Lorenzo Magnani, Walter Carnielli, and Claudio Pizzi, editors, *Model-Based Reasoning in Science and Technology: Abduction, Logic and Computational Discovery*, pp. 3-30, Berlin: Springer, 2010.

[66]It is open to question whether Nazi jurisprudence actually was a truth-maker of legal facts. But

question. If we could find a disciplined way to specify the converse domain of "x is true in y", we might think that we've advanced a goodly way in our quest to the concept of respects a degree of theoretical reputability. And with it we might bring some robust respectability to relativism.

> *Relativity without irrealism*: The semantic relativities of three-grade inconsistency robustness require no downgrading of realist presumptions.

For model-based science a central question is whether the idea that in the right circumstances, the truth of a proposition can be brought about by theoretical *stipulation* is an idea with legs. No one thinks that Doyle achieves a residency in London for Holmes by *theoretical* stipulation. No one thinks that juries bring about Spike's innocence and Ike's guilt by *theoretical* stipulation. But pure mathematics and abstractly model-based science are different. They are different enough to lend some encouragement to the current suggestion. There is no want of takers for the idea that in the abstract sciences the theorist is free to make things true. Not everyone likes this suggestion, needless to say. But it is wrong to say that it lacks for backers.

The present essay has been a promissory note in support of inconsistency robustness research programmes. After a final observation, I'm going to end with a further call for some further research. I want to propose that the idea of truth by stipulation in the abstract science be given renewed and sustained attention in the philosophy of science and semantics of mathematics.

The last word is this. If truth-making by stipulation is allowed in science, even for propositions known to be radically false on the ground, all the trouble we've so far canvassed regarding fiction recurs with a vengeance. That alone would motivate extending the reach of the IR-hypothesis to the model-based sciences and mathematics. If it worked, it would be stirring confirmation of the philosophical importance

---

suppose for a moment that it that it was. In it is a useful comparison with the Spike case. Everyone knows that Spike did the dirty deed even the jury that make it a legal fact that he didn't. This is upsetting. The legal fact is given behavioural priority over the non-legal fact. Spike walks instead of going to prison. Given the legal fact about him, no agent of government or anyone else can now proceed against Spike in ways that disconform to the legal fact of innocence. It is even worse the other way. Take the case of Ike. A vicious murder has occurred. It is a murder in which Ike lacks any and all involvement. But in due course a lawfully constituted and properly behaving jury made it a fact that Ike did it; and Ike spent the spent the rest of his life in federal prison. What would we say about this? We could change our minds about the truth-making wherewithal of jurisprudence hence the status of a respect in the way the world is, in the manner contemplated by LNC. But we could stick to our guns concerning the truth-making powers of fiction and theoretical science. This would be pretty impressive as such. Or we could stick to our guns about legal fact-making and concede that it can sometimes be a misfortune to give to legal facts the priority we customarily do. Seen that way, the holocaust case would be just the Ike case writ horrifically large.

of Hewitt's intriguing idea.

Acknowledgements

First and foremost, I warmly thank Carl Hewitt for the idea of inconsistency-robustness, and for several month's worth of instructive and encouraging e-correspondence. Frank Hong's sharp-eyed criticism of earlier drafts are also deserving of thanks, all the more so since Frank is just now finishing a BA in Honours Philosophy at UBC. I have also benefited from shrewd and generously detailed advice from this journal's anonymous referee and from four others whom Hewitt was kind enough to recruit for the project, whose reports were also passed onto the journal's editors. I am deeply grateful for their excellent advice. My recent thinking about fiction owes much to the influence of Shahid Rahman and his Lille Logic Group, as well as Alirio Rosales and Jillian Isenberg. My views of model-based science has benefited considerably by discussions with Lorenzo Magnani, Ahti-Veikko Pietarinen and Rosales. For the past twenty six years, Bryson Brown has been my "go-to-guy" in matters of paraconsistency, as has Graham Priest when true contradictions crop up on the order paper. Dov Gabbay and I have been talking about inconsistency-tolerance for nearly twenty-five years, and my own evolving views owe a great deal to that stimulus. For technical support, and all other things that matter to me, I am attached and immeasurably grateful to Carol Woods.

          Received 11 November 2013

www.ingramcontent.com/pod-product-compliance
Lightning Source LLC
Chambersburg PA
CBHW080940030426
42339CB00008B/463